Thomas Hartmann · Lenka Slavíková ·
Simon McCarthy
Editors

Nature-Based Flood Risk Management on Private Land

Disciplinary Perspectives on a Multidisciplinary Challenge

Editors
Thomas Hartmann
Landscape and Spatial Planning
Environmental Sciences Group
Wageningen, The Netherlands

Lenka Slavíková
Faculty of Social and Economic Studies
Jan Evangelista Purkyně University
Ústí nad Labem, Czech Republic

Simon McCarthy
Flood Hazard Research Centre
Middlesex University
London, UK

https://doi.org/10.1007/978-3-030-23842-1

This open access publication is based upon work from COST Action LAND4FLOOD - Natural Flood Retention on Private Land, supported by COST (European Cooperation in Science and Technology), www.cost.eu.

Foreword: Nature-Based Flood Risk Management

Extensive resources have been spent to foster the understanding of risks from flooding and to design and, where possible, implement risk reduction interventions, particularly in the last two decades. The benefits are visible in both science and society, given the widespread knowledge on FRM and the effects of implemented biophysical measures and policy instruments. However, flood disasters and risks remain one of the most demanding challenges from natural hazards worldwide. This calls for additional efforts, notwithstanding the inherent characteristic of risk that it may not be fully prevented.

Advancements in science and practice can build on the historical development of dealing with and preparing for flooding so far. After flood protection, the risk management paradigm has been promoted since the 1990s. The latter has led to, for example, an inclusion of exposure and vulnerability in the assessment of flood challenges, increased diversity of flood risk reduction interventions, a relative evaluation of risk instead of targets of absolute flood protection and greater attention to the societal processes of managing flood risks within the governance context. Knowledge and practices have profited from interdisciplinary and cross-sectoral collaboration, respectively, partly even from transdisciplinary co-design, co-deployment and co-evaluation. Hence, the question: what are the next steps? How to advance the relevance of future efforts in science and society to further reduce flood risks?

One possible way forward seems to tackle real-world flood issues more comprehensively. "Comprehensiveness" in this case means the degree of considering (all) aspects involved in FRM with their contextual, spatial and temporal features and interrelations. It does not assume that everything could and should be understood and managed. Instead, it raises the question where, among the wide range of aspects with their interrelations, may we lack understanding and management although they are relevant. This question addresses real-world phenomena and hence is ontological. Another possible way forward is to strengthen the conceptual and methodological ties between the description and explanation of individual aspects of FRM to better represent interdependencies. This way is driven by knowledge gaps and methodological limitations, which is thus epistemological.

Natural and social science systems' approaches trying to further conceptualise and operationalise complexity, dynamics and social construction could be of added value there. Neither way is exclusive of the other; they just represent two options for future efforts.

Land as a natural integral of atmosphere, geosphere, hydrosphere and biosphere is one key realm for flood risk generation and hence risk reduction. Land at the same time is the biophysical integral of natural and anthropogenic processes with the latter ranging from near-natural to technical interventions. And land represents immaterial societal meanings regarding the three pillars of sustainability with facilitating and regulating institutions such as property rights, planning designations and land markets. According to its central role, land has always played a part in FRM research and practice. Nevertheless, the focus changed under the above paradigms. While land was the main target for being kept free from inundation under the flood protection paradigm, land with its assets became the receptor for risk reduction under the current FRM paradigm. Furthermore, land is supposed to integrate an even wider range of natural and societal processes as the multiple aforementioned dimensions show.

In line with that, there are some indications that the consideration of land has not yet reached a degree of comprehensiveness that covers all relevant processes with their interdependencies. For example, there are recent efforts around the world, particularly in the USA and European Union, to study and apply NBS for flood retention. Respective measures are intended to stimulate or mimic natural processes while being integrated (e.g. agriculture) or separated from other land uses. Most of them interfere with the land, and some require significantly more land than hard engineering constructions, which they are often envisaged to complement rather than replace. The effects of altering natural processes on the land involve the full scope of the multidimensionality of land stated above. Hence, NBS raise the issue of comprehensiveness for the assessment of their (multiple) effects and implementation under the societal context. Hereby, both the assessment and implementation are likely to be framed by the diversity of natural and societal conditions and thus by the locality.

Taking such a comprehensive view on land makes FRM research and practice much more complicated than previous methods of flood protection and the current risk assessment and reduction. Research especially needs to consistently address (the same) land from a multidisciplinary view, drawing from an array of natural sciences, engineering sciences and humanities before it comes to an interdisciplinary and even transdisciplinary entanglement of concepts and methods towards an emergent knowledge. Among others, this can lead to a combined spatial and temporal assignment of various natural and societal land functions with their mobility. In addition to that, a FRM practice particularly faces the tasks of creating multi-actors' fora with the mandate of developing and implementing site-specific and even dynamically implemented measures and instruments to alter societal land functions together with landowners and users. While the prerequisites from individual disciplines and sectors are already well advanced, theoretical and methodological basics and practical experiences for tackling this comprehensiveness lag behind.

The challenges of properly addressing comprehensiveness are not limited to FRM; rather they are common for many thematic areas of environmental (risk) management. Current studies and discourses on, for example, nexus approaches and sustainable economies are indications for that. Insofar, FRM challenges are just one part of an overall task of tackling the complicacy of human–environment interrelations. At the same time, individual aspects involved in flood risk generation and risk reduction are mostly well understood; therefore, advancements for a more comprehensive FRM research and practice could become a thematic front runner also for other fields of environmental development and risk management.

This book, *Nature-Based Flood Risk Management on Private Land*, can be seen as a contribution towards a more comprehensive approach to land and land-dependent NBS. It addresses potentials and experiences with the practical application of NBS on private and partly public land from a wider view of various disciplines. Hereby, it exceeds a technical or impact assessment interest, which is undoubtedly the basis for any standardised design of respective measures and their transparent evaluation. Moreover, the aim of the book is to exploit existing projects as an initial step of medium-term studies.

After a conceptual setup, cases of fluvial flooding on different geographical scales are used for the analysis. All case sites are situated in headwaters or plains of river basins within central parts of the European Union. The approach is descriptive, involving, or at least reflecting, various disciplines. First, a disciplinary or interdisciplinary core team presents the NBS with their implementation. Second, comments from authors with other disciplinary backgrounds and not involved in the case widen the view and provide additional reflection. With this arrangement of the book, the cases serve as valuable illustrations. As a kind of report, they do not claim to present detailed evidence and explanation. In contrast, the cases become a point of departure for future studies to add depth of knowledge and practical experience.

Further analysis of these and other cases based on an inductive single and comparative case study design can then allow for the formulation of conceptual and methodological hypotheses, which may be empirically tested to gain evidence-based explanations. Hereby, and in addition to the societal relevance of the topic, there will likely be some room for the originality of results. The latter might range from development or adaptation of concepts and methods to frameworks for an integrated investigation of land in FRM. The outreach of this kind of research could then trigger innovation for FRM.

Following this, sequential approach could overcome a widespread understanding that the effectiveness of FRM mainly depends on the research and design of biophysical measures, while successful implementation is just a matter of negotiation, persuasion and maybe payment. On the contrary, it conceptualises the challenges of gaining more effectiveness through further entanglement of biophysical and societal aspects. This acknowledges that different biophysical processes of the land for flood generation and the societal meanings of this land according to the institutional context are seen as extensively interwoven.

The book remains humble in this respect. And this is one of its strengths since the scientific prerequisites for further ambition are not yet sufficient. Another merit is the rich empirical basis and the multidisciplinary reflection, which are highly informative and set the scene for a broader discussion of NBS, the implementation mechanisms and the societal context referring to the accessibility of the required land. Presentation of the cases remains predominantly neutral to avoid bias regarding simple and general pros and cons for certain types of measures. This openness inspires further discussions without promoting NBS as the sole future strategy.

Compared to some previous researches, the cases sharpen the perception of already known and uncover new barriers and enablers for the implementation of land-based flood risk reduction measures. For instance, the engagement of individual owners of huge farms has not been taken into account in earlier works on government-led vertical and community-based horizontal FRM strategies. At the same time, innovative planning and funding instruments are critically discussed, even admitting the higher degree of effectiveness of traditional vertical instruments. Of course, most of the cases are based on well-known natural-based solutions such as decentralised water retention and space for the river. Nevertheless, the involvement of property rights focusses on intersectoral coordination processes and others enhance the comprehensiveness of framing the FRM problems.

Overall, approaching land with NBS in the manner of the book *Nature-Based Flood Risk Management on Private Land* facilitates a move towards a more comprehensive and integrated manner of tackling flood issues. This poses valuable new questions for research and practice. How far the answers to these questions will support a higher effectiveness of flood risk reduction cannot be anticipated at this stage. Maybe, also limited land accessibility defines the tolerability of risk from a benefit–cost point of view since willingness to act and pay also determines the maximum benefits. To explore this in future studies and real-world cases with the comprehensiveness laid out in the book appears to be a timely and well-justified ambition.

Dresden, Germany

Jochen Schanze
Department Environmental Science
and Risk Management, Technische
Universität Dresden/DRESDEN-concept

Leibniz Institute for Ecological Regional Development
jochen.schanze@tu-dresden.de

Foreword: The Network LAND4FLOOD

In 2013, the Akademie für Raumforschung und Landesplanung (ARL) supported a summer school in Utrecht (Netherlands) organized by Thomas Hartmann and Tejo Spit on Sustainable Governance of Land and Water. The participants of the summer school were Ph.D. students and professors—and during that week in Utrecht, the participants had a chance to explore challenges of water and land governance. This was the real beginning of a future network that resulted in an exceptionally stable and productive cooperation. In 2014, the Czech team (Jiřina Jílková and Lenka Slavíková) responded to a project call from the Czech Ministry of Education within the LE-EUPRO II programme for networking projects and were successful. The aim of the project called CrossFlood (cross-border flood risk management) was the establishment of an interdisciplinary network of institutions and experts who share a focus on cross-border management of flood risks called floodland. The expertise of Elbe, Rhine and Danube river basin management authorities in governance and technical solutions in water and land management has been critically evaluated, and innovative solutions for complex planning approached have been discussed in accordance with the shift from flood protection to wider flood risk governance. The idea for a future COST Action—a special programme to fund academic networks in European Union—was formulated: in order to deal with flooding, the use of (private) land for flood water retention needs to be addressed.

In this way, the small Czech network project became a success story. Within its lifetime, the group of colleagues from different disciplines—including hydrology, spatial planning, economics, law and others—and different countries prepared the proposal for a wider network on the relation between land and flood risk management (FRM). Previously, this topic had barely been addressed in the academic debate, and the network had the ambition to put this topic on the academic and political agenda. In 2017, the proposal was approved and network Land4Flood (Natural Flood Retention on Private Land) started. The smooth transfer from the CroosFlood project to LAND4FLOOD COST Action is well captured in the joint special issue of the *Journal of Flood Risk Management* on "Land for Flood Risk

Management: A catchment-wide and multi-level perspective". The group continues to grow—now including academics and stakeholders from more than 30 countries—and it represents the important baseline for the future evolution of collaborative projects, publications and events.

This book on *Nature-Based Flood Risk Management on Private Land* demonstrates the specific character of the network—an interdisciplinary group, grounded in the field, presenting original perspectives from different contexts and at different scales. The book is organized into four parts. Part I sets up the overall framework related to the debate about nature-based solutions (NBS). The following parts are organized according to the scale of measures examined—small, medium-sized and large-scale measures. This format gives the reader a clear orientation and opportunity to compare perspectives.

The paper by Hartmann, Slavíková and McCarthy critically assesses how water management is currently changing its approach to floods in favour of more NBS. The authors mention the pitfalls and problems related to NBS and the specific differences in comparison with grey infrastructure measures: they require more land than traditional measures, and this land is often owned by private landowners. Hartmann et al. call the land a critical factor for determining the implementation of NBS. Land use—both urban and agricultural—has changed immensely over the last decades. These processes include a more intense use of urban areas, the increasing use of heavy machines and a change in the production structure, including the elimination of buffer zones on agricultural land. Compacted soil has decreased ability to hold water, which, in turn, speeds up the flow of water on the soil's surface. This increases the volume and speed of water flowing into waterways and heightens the risk of flooding. The current agricultural policy does not aim to stop this development. Also in the urban realm, land management is not yet considered an essential element of FRM. This book contributes to elaborating on how land matters for FRM in general and nature-based FRM in particular.

Even interventions on small scales may have big impacts on the larger scale, but knowledge of these connections is limited (O'Connell 2007). The authors mention that research has not yet produced conclusive evidence on whether local NBS can have an effect on large-scale flooding. Why are NBS favoured in the current policy on FRM? This book sets out to explore this question and searches for examples. This proves to be an interdisciplinary endeavour. Therefore, a special format for the book has been chosen: it is based on case studies and subsequent commentary solicited from different disciplinary perspectives.

Raska, *Slavíková* and *Sheehan* discuss the issue of scale in NBS for FRM. They derive the importance of the scale from the dichotomy between hydro-ecological and institutional approaches. They explain the scale in interdisciplinary perspective using various examples from history, geosciences and social sciences, often in historical perspective. They introduce both an epistemological and ontological view of scale. Their approach introduces a theoretical and a philosophical dimension to the book. However, their concept has very practical implications related to upscaling or downscaling decisions with the primary motivation to augment resilience. They analyse fragmentation of land and the mismatch in socio-ecologic systems as the real problems of land management. They conclude by emphasizing the need to realign the scales of flood risk measures.

Part II deals with small-scale property solutions. Case studies from three countries are presented here—Poland, Czech Republic and Flanders (Belgium). *Matczak*, *Takacz* and *Goździk* present the small retention programmes in the Polish forests. Slavíková and Raska present in their paper a different case: measures taken on privately owned agricultural land. The study presents an exceptional case for the post-socialist planning context where the engagement of private actors is almost forgotten after 40 years of collective management of the land. The authors call this approach an environmentally concerned philanthropy and leave the question of upscaling open.

Part III of the book is devoted to medium-sized infrastructure. The accent shifts to measures like urban wetland restoration in floodplains. The paper by *Macháč* and *Louda*, and the related commentaries by *Gutman* and *Pohl*, presents the case of the Czech city Pilsen. The other case (presented and commented in three papers) is the Blauzone Rheintal—example of a measure encompassing a whole catchment.

Large-scale catchment solutions are presented in Part IV of the book. The cases include the relocation of dikes (river landscape Elbe–Brandenburg in Germany and Oekense Beek in the Netherlands). *Warner* and *Damm* look at different measures from different perspectives—hydrology, finance, governance, property rights. In comparison with small-scale measures, large projects bring greater complexity in relation to financing, ownership and project initiator interactions. Also, *Kaufmann* and *Wiering* proof that from the flood retention plan to making this plan a reality a lot of land needs to be covered. These examples from the book illustrate the variety of approaches to nature-based FRM from different disciplinary angles.

The pressing question in this context is: how do initiatives go from isolated positive cases to a legitimate, outspread practice, or, in other words, how to scale-up successful practices? There is no simple answer. The authors and commentators have clearly shown that, if we want to achieve flood retention relevant from a hydrological point of view, a couple of aspects need to be considered:

- The catchment dimension is relevant, and at the same time, it needs to be broken down into a set of locally specific measures.
- An intersectoral and thus interdisciplinary coordination between flood management and spatial planning is essential for nature-based flood risk management.
- Public funding remains utterly important to realise flood risk management. Market and commons-led initiatives are not yet a viable and only exceptional alternative for a central government-coordinated approach.
- Nature-based flood risk management asks for more inter- and transdisciplinary communication and research.

The idea of NBS is discussed in all contributions, whereas the tone of the book is not about NBS as a superior strategy; it is rather about the challenges that are precipitated by the inevitable interaction with the owners often privately owned land within a river basin. The diverse from multiple disciplines in this volume contributions show that in order to find alternative, more efficient, and effective ways to deal with floods is indeed a multidisciplinary challenge.

Ústí nad Labem, Czech Republic

Jiřina Jílková
Faculty of Social and Economic
Studies/Institute for Economic
and Environmental Policy (IEEP)
Jan Evangelista Purkyně University
jirina.jilkova@ujep.cz

Contents

About the Editors

Dr. Thomas Hartmann is Associate Professor at the Landscape and Spatial Planning Group of Wageningen University. He combines in his research an engineering perspective on environmental science with a socio-political perspective on FRM and land policies. He is Vice-Chair of the Land4Flood COST Action, Speaker of the advisory board of the German Flood Competence Centre and Active Member of the OECD Water Governance Initiative. He is also Vice-President of the International Academic Association on Planning, Law and Property Rights (PLPR).

Lenka Slavíková graduated from the University of Economics, Prague (public economics and policy). Currently, she serves as Associate Professor at the Faculty of Social and Economic Studies, J. E. Purkyně University in Ustí nad Labem. Her long-term interest is in water and biodiversity governance with the focus on Central and Eastern European Countries. She investigates flood risk perception of different actors and financial instruments for flood recovery and mitigation.

Dr. Simon McCarthy with a background in commercial social research, undertakes teaching, training and academic social research in the Flood Hazard Research Centre at Middlesex University London. His research interests focus on the role of both public and professional social contexts in decision-making, communication of risk and uncertainty and approaches to participatory interaction in flood risk and water management. He is appointed as Member of the Thematic Advisory Group on flood and coastal erosion risk management research and development for England and Wales; Department of Environment, Food and Rural Affairs, Environment Agency and Natural Resources Wales.

Part I
Introduction

Chapter 1
Nature-Based Solutions in Flood Risk Management

Thomas Hartmann, Lenka Slavíková and Simon McCarthy

Floods are among the most expensive natural disasters (Munich Re 2014). The Intergovernmental Panel on Climate Change (IPCC) states—with "high confidence"—that damages incurred by water-related risks continue to increase in Europe (IPCC 2014) due to changing hydro-meteorological conditions. Also landslides or avalanches are among the consequences of these developments. But not only changing environmental conditions, but also intensification of land and water use, contribute to increasing risks. In particular, cities are increasingly vulnerable to such events—as recent flash floods in Central Europe have illustrated in summer of 2016.

In recent decades, water management has been changing its approach: although technical and engineering methods and measures are still prevailing in many subfields of water management, nature-based solutions (NBS) are growing more popular. However, the frequency, variability and scale of their implementation vary throughout Europe.

Nature-based solutions are "actions which are: (1) inspired by, (2) supported by or (3) copied from nature" (European Commission 2015, p. 5). Such solutions for risk reduction and adaptation in river catchments involve, for example, Natural Water

T. Hartmann (✉)
Faculty of Environmental Sciences, Wageningen University & Research, Wageningen, The Netherlands
e-mail: thomas.hartmann@wur.nl

Faculty of Social and Economic Studies, J. E. Purkyně University in Ústí nad Labem, Usti nad Labem, Czech Republic

L. Slavíková
Faculty of Social and Economic Studies, Institute for Economic and Environmental Policy (IEEP), J. E. Purkyně University in Ústí nad Labem, Usti nad Labem, Czech Republic
e-mail: lenka.slavikova@ujep.cz

S. McCarthy
School of Science and Technology, Flood Hazard Research Centre, Middlesex University, London, UK
e-mail: S.McCarthy@mdx.ac.uk

T. Hartmann et al. (eds.), *Nature-Based Flood Risk Management on Private Land*,
https://doi.org/10.1007/978-3-030-23842-1_1

Retention Measures (NWRM), space for the rivers, or measures for resilient cities (i.e., green infrastructure in cities, green roofs, decentralized rainwater management). These solutions are also referred to as "green and blue infrastructure". Typically, such measures bring multiple benefits to people and social systems—they can, for example, not only reduce flood risks but are able to simultaneously improve the quality of life, reduce heat and dust, enrich biodiversity, etc. Nature-based solutions to water-related risks cannot entirely substitute for traditional measures such as flood pathway and receptor approaches, both structural and behavioral (e.g., flood walls, flood warnings), but their potential value for risk reduction and adaptation has been recognised (European Commission 2015).

Natural flood management (NFM) and natural water retention measures (NWRM) are also types of NBS; NFM includes measures that "alter, restore or use landscape features to manage flood risk" (Holstead et al. 2015); NWRM include (1) interception (retaining water in and on plants), (2) increased plant transpiration, (3) improved soil infiltration, (4) ponds and wetlands, and (5) reconnecting the floodplain. These measures have the potential to reduce extremes in the flow discharge and thus help to level out extremes. Positive effects can include a beneficial impact on ecological issues (i.e., nutrition retention), agriculture (irrigation) or tourism. Natural water retention measures can be combined with other aims of water management—most notably with water quality (Morris et al. 2014)—but also with agriculture, tourism or ecology (Posthumus et al. 2008; Calder 2005; Biswas 2004). But the ancillary benefits of NWRM, the compatibility of different purposes, and the cumulative effects have hardly been researched, as an initiative led by the European DG Environment on NWRM has shown (www.nwrm.eu).

Also floodplain restoration in general can be considered a nature-based solution that mitigates water-related risks (European Commission 2015, p. 12). The concept of making space for the rivers was first introduced by the Dutch Government as a reaction to the major floods in the 1990s (Greiving 2002). "Space for the rivers" summarizes a paradigm that moves from the ideology of defending against floods and "keeping the water out" to an ideology of managing floods and asking citizens to "make space for water" (Johnson and Priest 2008, p. 513). Besides preventing flood damage, space for the rivers can also have beneficial effects for the environment (Moss and Monstadt 2008). The European Commission has already affirmed in 1999 in the European Spatial Development Programme that river works and urban development in the floodplains accelerated flood risk (article 319). In addition "restoring degraded terrestrial ecosystems, such as grasslands, arable land and forests, as well as former industrial and brownfield sites by using nature-based solutions also can deliver a variety of benefits, including improved water quality, carbon sequestration, and attractive landscapes" (European Commission 2015, p. 18). At the time, policy initiatives to restore floodplains are limited to a few forerunners (Moss and Monstadt 2008, p. 64). Still today, implementation of space for the rivers is hampered by the lack of available rights in land (i.e., land use and land ownership) (Hartmann 2012).

A related concept to NBS is called "Payments for Environmental Services (PES)" (sometimes Payments for Ecosystem Services). These measures go beyond NBS. They involve "redistributive mechanisms between different social groups" (Kumar and Muradian 2009, p. 1) that aim to take into account environmental services provided by one party for the service of others (Kumar and Muradian 2009, p. 8). Watershed developments are an application of PES schemes in developing countries, in particular in India (Kerr 2002). These projects seek "to optimize the use of natural resources for conservation, productivity, and poverty alleviation" (Kerr 2007). So, PES includes a conceptual approach.

So, there are related concepts in flood risk management (FRM), which can be summarised under the term "nature-based solutions". The current trend towards NBS has emerged as traditional ("grey") infrastructure (such as dikes and dams) has been questioned increasingly and rejected by some scholars while actual or assumed benefits of NBS have been emphasised. Grey infrastructure is usually very specialised—specifically designed to solve one particular issue (i.e., defend the centennial flood event). These measures are generally not versatile enough to address different issues; hence changing environmental conditions present a challenge to these types of solutions. In theoretical terms, grey infrastructure is often prone to technological lock-in situations (Arthur 1989; Thompson 2008). However, changing societal needs and dynamic nature (i.e., climate change) lead to a need for more multi-functional and flexible solutions. Nature-based solutions are assumed to be much more adequate for multi-purpose use than traditional grey infrastructure. Nature-based solutions cannot replace grey infrastructure but rather be integrated so that more traditional methods of management are complemented or enhanced by using nature (European Commission 2015, p. 12). Nature-based solutions are suggested by the EU as a complementary and sustainable way of addressing "a variety of environmental, social and economic challenges" (European Commission 2015, p. 5). In the current research funding landscape, NBS seem to be regarded as panacea for many environmental issues.

Nonetheless, some pitfalls and problems related to NBS need to be considered. One of the issues is the great uncertainty of the effects of many NBS. The effects are difficult to quantify, and therefore they defy traditional methods to assess and justify measures. Usually, for example, dikes are justified via a positive cost-benefit ratio, which confirms that the costs of building such structures will pay off in terms of prevented damage. But if NBS cannot be justified in this way, their realisation—in particular when it means to intervene in private property rights or to spend a lot of public money—can come into question. This becomes crucial because NBS need more land than grey infrastructure, as we will discuss below.

Nature-based solutions have two interrelated issues in common: first, basically most such measures require more land than traditional grey infrastructure. A dike against inundations, for example, is much more land thrifty than a retention area. Within retention areas, controlled retention areas are far more effective, but alluvial forests are much more valuable in terms of their ecological benefit. Although this oversimplifies the matter; as a general rule, the more nature-based a solution is, the higher its demand for land. Second, the land that NBS need is often owned by private landowners rather than public stakeholders. These measures raise conflicts over land (Van Straalen et al. 2018).

Ultimately, land is the critical factor that determines whether NBS can be implemented to deal with water-related risks. Land is an essential and inevitable ingredient for the implementation of NBS to mitigate and adapt to water-related risks. Making this land available and persuading land users to implement the measures are thus two key challenges for implementing measures to mitigate or adapt to water-related risks. Usually, flood risk management deals first with technical and hydrological issues before addressing land management. Implementation of flood risk management is often hampered by the lack of land management approaches. Land users are often regarded as mere recipients of water management, not as key stakeholders. Most existing research initiatives on water-related risks focus on technical or hydrological aspects, forecasting, disaster management, or institutional governance aspects. Approaches for collaborating with private land users to realize risk reduction and adaptation measures on private land are lacking in theory and practice. Although there are many case studies and much experience on NBS on the small scale, evidence on the effectiveness and efficiency of nature based-solutions on a large scale is lacking. "There is a clear need to compile a more comprehensive evidence base on the social, economic and environmental effectiveness of possible NBS, including a comparison with more traditional solutions" (European Commission 2015, p. 21). If land management for NBS is not properly addressed and scaled up to the level of the catchment (or aquifer), NBS for FRM remain ineffective and inefficient.

In conclusion, nature-based solutions are favoured in FRM. These measures require more—and mostly privately owned—land, and more diverse stakeholder involvement than traditional (grey) engineering approaches. This also implies that there are challenges related to different disciplines. Flood risk management with NBS is an issue not only of technical expertise, but it asks for land-use planning, economics, property rights, sociology, landscape planning, ecology, hydrology, agriculture and other disciplines to cope with the challenges of implementing them. Ultimately, nature-based FRM is an inter and transdisciplinary endeavor. This is why this volume is addressing the various disciplinary aspects of NBS in FRM on private land.

Two related research questions are therefore discussed:

1. What are the challenges of NBS from various disciplinary angles?
2. How can a inter and transdisciplinary approach to nature-based FRM help dealing with these challenges?

This requires a special format. Therefore, cases are presented that develop, demonstrate or deploy innovative systemic and yet locally attuned NBS (i.e., green and blue infrastructure and ecosystem-based management approaches, in rural and urban areas). The role of land management and spatial planning is described as well as the involvement of other stakeholders. All cases reflect on the multi-benefit of the measures, such as impacts on landscape, local communities and cultural acceptance as well as co-benefits such as biodiversity conservation/enhancement, more sustainable local livelihoods, human health and well-being, climate change mitigation, etc. Barriers related to the social and cultural acceptance and policy regulatory frameworks will be discussed as well as a reflection on the replication of innovative investment strategies elsewhere.

A special feature of the book is that each case study is discussed from at least two different disciplinary perspectives. So, the main body of the book comprises two kinds of contributions: main contributions outline a case study of NBS. The contributions will address a description of the problem (why some action started), the relevant contextual variables (biophysical environment, socioeconomic conditions, and institutional arrangement), the main actors and their interactions (with focus on conflicts, cooperation, and social capital creation), an outline of how the action pursued/implemented (i.e., procedural aspect and governance). These main contributions are then complemented by shorter commentaries, where authors discuss the presented solutions in the case through the lens of particular disciplines. The commentaries are brief academic reflections that critically highlight which specific aspects are of significance from a certain disciplinary angle.

Acknowledgements Open access of this chapter is funded by COST Action No. CA16209 Natural flood retention on private land, LAND4FLOOD (www.land4flood.eu), supported by COST (European Cooperation in Science and Technology).

References

Arthur WB (1989) Competing technologies increasing returns and lock-in by historical events. Econ J 99:116–131

Biswas AK (2004) Integrated water resources management: a reassessment. Water Int 29(2):248–256

Calder IR (2005) Blue revolution. Integrated land and water resources management, 2nd edn. Earthscan, London

European Commission (2015) Towards an EU Research and Innovation policy agenda for Nature-Based Solutions & Re-Naturing Cities. Final report of the Horizon 2020 Expert Group on 'Nature-Based Solutions and Re-Naturing Cities'. European Commission, Brussels

Greiving S (2002) Räumliche Planung und Risiko. Gerling-Akademie Verlag, München

Hartmann T (2012) Land policy for German rivers. Making space for the rivers. In: Van Buuren A, Edelenbos J (eds) Warner JF. Making space for the river. Governance experiences with multifunctional river planning in the US and Europe. IWA Publishing, London, pp 121–133

Holstead KL et al (2015) Natural flood management from the farmer's perspective. Criteria that affect uptake. J Flood Risk Manag 10(2):205–218

IPCC (2014) Climate change 2014: impacts, adaptation, and vulnerability. IPCC WGII AR5, Part B: regional aspects, pp 1267–1327. http://www.ipcc.ch/report/ar5/wg2/. Accessed 22 July 2014

Johnson CL, Priest SJ (2008) Flood risk management in England. A changing landscape of risk responsibility. Int J Water Resour Dev 24(4):513–525

Kerr J (2002) Watershed development, environmental services, and poverty alleviation in India. World Dev 30(8):1387–1400

Kerr J (2007) Watershed management. Lessons from common property theory. Int J Commons 1(1):89–109

Kumar P, Muradian R (2009) Payment for ecosystem services. Oxford University Press, New Delhi, New York

Morris J, Beedell J, Hess T (2014) Mobilising flood risk management services from rural land: principles and practice. J Flood Risk Manag 9(1):50–68

Moss T, Monstadt J (2008) Restoring floodplains in Europe. Policy contexts and project experiences. IWA Publishing, London

Munich Re (2014) Schadenereignisse weltweit 1980–2013. 10 teuerste Überschwemmungen für die Gesamtwirtschaft. Munich Re. Available via DIALOG. https://www.munichre.com/…/Significant-Natural-Catastrophes/.pdf. Accessed 1 Aug 2014

Posthumus H et al (2008) Agricultural land use and flood risk management: engaging with stakeholders in North Yorkshire. Agric Water Manag 95(7):787–798

Thompson M (2008) Democratic governance, technological change and globalisation. Economic and Social Research Council (ESRC). Available via DIALOG. http://cultural-theory-action-research.wikispaces.com/file/view/Michael+Thompson+Lisbon+Report+Democratic+Governance+Technological+Change+and+Globalisatio+%282%29.pdf. Accessed 19 Oct 2012

Van Straalen F, Hartmann T, Sheehan J (eds) (2018) Property rights and climate change. Land-use under changing environmental conditions. Routledge, Abington

Thomas Hartmann is Associate Professor at the Landscape and Spatial Planning Group of Wageningen University. He combines in his research an engineering perspective on environmental science with a socio-political perspective on FRM and land policies. He is vice-chair of the LAND4FLOOD Cost Action, speaker of the advisory board of the German Flood Competence Centre and active member of the OECD Water Governance Initiative. He is also Vice-President of the International Academic Association on Planning, Law, and Property Rights (PLPR).

Lenka Slavíková graduated from University of Economics, Prague (public economics and policy). Currently, she serves as the Associate Professor at the Faculty of Social and Economic Studies J. E. Purkyně University in Ustí nad Labem. Her long-term interest lies in water and biodiversity governance with the focus on Central and Eastern European Countries. She investigates flood risk perception of different actors and financial instruments for flood recovery and risk reduction.

Simon McCarthy with a background in commercial social research, undertakes teaching, training and academic social research in the Flood Hazard Research Centre at Middlesex University London. His research interests focus on the role of both public and professional social contexts in decision making, communication of risk and uncertainty and approaches to participatory interaction in flood risk and water management. Simon is appointed member of the Thematic Advisory Group on flood and coastal erosion risk management research and development for England and Wales; Department of Environment, Food & Rural Affairs, Environment Agency and Natural Resources Wales.

Chapter 2
Scale in Nature-Based Solutions for Flood Risk Management

Pavel Raška, Lenka Slavíková and John Sheehan

Introduction

Nature-based solutions (NBS) have recently spread to the flood risk management (FRM) agenda as potentially efficient and sustainable measures to reduce the susceptibility to and impacts of various kinds of floods, including riverine floods, flash floods and storm surges. In this context, some authors urge for a deeper understanding of the hydro-ecological effects of NBS on different scales and for the diversity of environmental conditions (Schanze 2017; Nesshöver et al. 2017). At the same time, suitable institutional scales for effective implementation of NBS in different legal settings must also be set up. If these two approaches, hydro-ecological and institutional, shall work together and facilitate the implementation of NBS in FRM, it is crucial to understand the different meanings of scale across the disciplines that are involved in FRM.

Only few concepts used in both social and environmental sciences have merited such critical reflection and extreme controversies during the past decades as the issue of scale (Marston et al. 2005). At the same time, similarly few concepts may affect the effectiveness and efficiency of the environmental management as much

P. Raška (✉)
Department of Geography, Faculty of Science, J. E. Purkyně University in Ústí nad Labem, Usti nad Labem, Czech Republic
e-mail: pavel.raska@ujep.cz

L. Slavíková · J. Sheehan
Faculty of Social and Economic Studies, Institute for Economic and Environmental Policy (IEEP), J. E. Purkyně University in Ústí nad Labem, Usti nad Labem, Czech Republic
e-mail: lenka.slavikova@ujep.cz

J. Sheehan
e-mail: John.Sheehan@uts.edu.au

J. Sheehan
Faculty of Society and Design, Bond University, Gold Coast, Australia

T. Hartmann et al. (eds.), *Nature-Based Flood Risk Management on Private Land*,
https://doi.org/10.1007/978-3-030-23842-1_2

as scale does (Cumming et al. 2006). The aim of this chapter is to introduce the roots of various conceptualizations of scale, the way they are encountered, and the implications different views of scale may have on the use of NBS in FRM.

The literature currently available on issues of scale is generally grounded in geography and spatial science, but also crosses into environmental management, economics and other disciplines such as law. In FRM, scale has been widely discussed in the context of a range of issues, starting with a priori flood risk assessment and pointing to differential methodologies, use of flood risk assessment and uncertainties across scales (de Moel et al. 2015; Gusyev et al. 2016), extending to issues of restoration efforts aimed at improvement of overall river quality (Muhar et al. 2016). Among the studies, the theoretical work on scale in complex socio-ecological systems and on scale mismatches has fundamental implications for effectiveness and efficiency of FRM using the NBS—this will be mainly addressed in this chapter.

Scale in complex socio-ecological systems is directly impacted by the specificity of a chosen nature-based solution in FRM which, in turn, is similarly impacted by the particularity of the overarching property laws and especially, local land uses and property ownership patterns. Specificity of a FRM solution and particularity in local land tenure are not unexpected but rather indispensable contemporaneous fundamentals when determining how suitable a scale is. This can help to avoid scale mismatches.

Therefore, the ambition of this chapter is not to summarize the knowledge on scale nor to develop new concepts of the term. Rather—since the chapter should serve as a prelude to case studies presented later on—our aim is to provide a comprehensible essay on scale, explain the fundamentals of its nature and discuss its implications for NBS in FRM. We will endeavour to follow an interdisciplinary perspective and to make use of various examples from history, geosciences and social sciences to introduce readers to complexities of scale. First, we will discuss the two views of scale, which are the epistemic and ontological ones, and will point out that scale does not only relate to space but also to time, and that a change in any of these dimensions may incur a change in the other one. Second, drawing from these considerations, we will then discuss issues of downscaling and upscaling, as these are crucial in turning the experimental and/or theoretical research results into practice. Finally, we will summarize major constraints one encounters in managing socio-ecological systems such as NBS in FRM and resulting from the different conceptualizations of scale.

Nature of Scale in Nature-Based Solutions

Epistemic View of Scale

Practitioners seeking for a definition of scale in traditional dictionaries may remain frustrated in their effort. The online version of Merriam-Webster dictionary, for instance, defines scale in its conventional connection to weights and therefore denotes its original purpose to compare the Newtonian properties of the material world. It will be shown later on that such understanding of the scale may be relevant for many bioecological and biophysical studies that make use of the so-called geometrical scale. However, while absolutely necessary in our daily life (take just health care or shopping as examples), this perspective reduces the reality, because it assumes that all things may be compared in an objective manner. But what we already know is that some realities are rather relative and—in a social domain—may be differently evaluated by people and institutions (not) involved. To describe the reality in such manner, we usually refer to human-geographical or sociological scale as the representation of reality (Gregory et al. 2009; Mayhew 2015).

From this perspective, scale is a fundamental measure through which we perceive and understand the world around us (Lynch 1960). Imagine a small child walking through the crowded town and perceiving the surroundings from an eye-height of less than a meter, for instance. As this child escapes a jumble of legs surrounding it, it sees the far horizons of streets with buildings so high that it must bend backwards to see their facades and to orient itself within the known and unknown world. The adults, in turn, are well oriented all the time, but it may easily happen that they lose sight of their children or even bump into them. Globally, we recall the notion of intangible indigenous cultural knowledge (Office of Environment and Heritage 2017) wherein some savanna tribes in sub-Saharan Africa, and in particular Australian aboriginal peoples, are remarkably well orientated in their vast African or Australian grasslands. Alternatively, those Indigenes would almost certainly feel very disorientated in the dense primeval highland forests of Papua-New Guinea.

What this basic distinction between geometrical and geographical or sociological space illustrates is the epistemic view on scale, both in terms of practice (shaping the world around us) and of the research communication (targeting our research and finding a common language). Therefore, the epistemic dimension of scale expresses its function in terms of our orientation in the world, but, at the same time, it denotes how the space around us is socially constructed and how it may be consumed and reproduced (Brenner 2001). Further examples of such relativity of scale may be found in history. Until the last third of the 20th century, many historians were extensively addressing the landscape histories by converting the old units to the metric system. It was a well-known Russian historian A. J. Gurevic (1985), however, who pointed out the social construction of these old units early on. For instance, the medieval measures of fields in England were expressed as an expanse that may be ploughed by eight oxen in a season, and in Czech lands, the area of meadows was expressed as the number of drays necessary to transport all the yielded grass. Neither of these units was normalized, and both differed across regions significantly. Yet, both of them provided people with clear impression of the size, resolution and dimension since these units related to peoples' daily practices (see also Alfonso 2007).

A discussion of how various factors and agents interact at various scales is forthcoming. However, the very first implication we may present now for implementation of NBS in FRM is this: different stakeholders scale a certain space differently. In a large catchment, for instance, if river basin authorities were to design a flood retention measure on a few hectares in order to reduce the riverine floods, this may seem too extensive for a landowner if this comprises almost the whole of his/her land. Meanwhile it would only represent a small piece of the intended effort for the authorities. For NGOs that helped during the negotiation process, it may be a great story of success. In contrast, the landowner who designed the measures on his/her private land may be well satisfied with their effect, while the river basin authorities may not consider such measures as relevant to their catchment-scale effort (Slavíková and Raška—Part II, Chap. 2; the role of scale in multilevel management is generally discussed by Termeer et al. 2010).

Until now, we were talking about spatial scale, but issues of scale similarly relate to time. Much of the social construction of time is constrained by the level of our knowledge. Following this way of argumentation we may notice, for example, that with growing number of available data sources towards present days, the historical periods established by historians are becoming shorter. Historians of modernity may also exploit the fascinating potential of oral history, thus gaining a very variegated view of the near past as compared to the ancient times; so their descriptions of the past (with some exceptions) are becoming more detailed. Furthermore, there is certainly not a single definition of the historical periods across the continents, or even the regions, that also denotes how time is considered in various cultures. Thus, Green (1995), for instance, has shown that eras of the world's history tend to reflect our priorities and values, while raising political questions and concerns.

What makes all these time-scale considerations relevant to our topic is, that space and time do not only form a duality but a rather intertwined and complex time-space complexities (Hägerstrand 1970). This is because changing the time dimension may alter how things are organized and interdependent in a spatial dimension, and vice versa. Harvey (1989) introduced the apt term of time-space compression to describe how space was reduced with Man's increasing activity radius throughout history. This also implied the growing speed of social processes. For instance, in spite of various feedbacks and control mechanisms, some institutions or individuals currently possess the competence and ability to significantly affect our world in only a few short minutes. On the other hand, there still are extensive communities that may have almost negligible influence on institutional structures they live in, or their effort would only produce slow or delayed structural changes that will not immediately address the critical issues faced by these communities (cf., Giddens 1984).

The implication of the epistemic view of the time scale (chronology) and of the time-space complexity for FRM is twofold. First, as the perceived time dimension may lead (and often does) to categorizing landscape processes in terms of stability and instability, it may result in misperceptions of fluvial hazards (Schumm 1994) in turn. According to this view, some disturbing processes (e.g., floods) are wrongly perceived as solely non-natural and rapid, while certain human-induced changes may be perceived as everlasting. In other cases, the longevity (see further text) may be expected as a fundamental property of man-made FRM interventions (e.g., the small-scale NBS discussed in Matczak—Part II, Chap. 1). Such a perception of stability may then influence the decision about their implementation if these measures are finally evaluated as unstable. Second, the epistemic view of time-space complexities points to the increasing variability of speed and extent with and to which institutions and individuals may make their decisions about flood risk and land management.

Ontological View of Scale

The notion of subjectivity in perception of space and of relativity of scale inevitably begs the question whether any kind of scale-based organization of reality independent of our minds exists in bioecological/biophysical and social systems. Or, being consistent with our terminology above, how relevant is the ontological view of scale? In scholarly literature, this question is usually addressed in two directions (Herod 2003). First, whether—generally speaking—the links through which objects and processes are related change across scales, and second, whether any ontological scales exist, such as global and local, independent of our construction of space.

Some scientists would probably argue that ontological scale is more characteristic of the bioecological and biophysical domain (e.g., Cumming et al. 2006). Although some of the sole physical objects and features may display similarities across scales (well explained by the fractal theory), links among such objects and features may be subject to variations when seen through changing time-scale prisms, in fact. Such variations are widely discussed in biology, ecology or geomorphology. In their seminal paper, Schumm and Lichty (1965) addressed interdependencies of variables in the river drainage systems and showed how dependent variables may become independent when seen in different time frames, and vice versa. In other words, the causation among variables may differ within the time and spatial scale. Therefore, rather than being simply hierarchical, various features and processes may exhibit differently set continuities (Marston et al. 2005) and feedbacks across space and time.

Another illustration of variations in the organization of bioecological and biophysical features stems from their adaptability. While some features may display a high potential for changes, others—and viewed on different scales—evince inertia caused by negative feedbacks, or their modifications are constrained by physical, chemical and biological laws that do not allow for a significant change. These differences, characterized as longevity of socio-ecological systems and subsystems (Costanza and Folke 1996, p. 21), should frame our evaluation of sustainability of environmental management measures, including FRM.

The second question, whether ontological categories of scale such as global and local exist, is subject to intense debates (Herod 2003; Marston et al. 2005). In any case, both social sciences and geoscience tradition would probably admit that what makes an agent or a process global is not a priori its size but largely depends on the strength of its causal links to influences on the global system (Wilbanks and Kates 1999). Typically, extreme increases in the quantities of greenhouse gases produced by only a couple of countries are considered the main factor contributing to global climate change, including growing extremes in weather pattern, rising sea levels and resulting social impacts (Sheehan 2014). The existence of social and economic ties established by a strong or innovative leader or a firm can have the capacity to influence a global production network and/or diffusion of innovation in such a way that local economies may be significantly affected (Ernst and Kim 2002). Nevertheless, neither of these examples must imply the uniformity of impact caused on a local scale. This justifies the current study of regional environmental change or regional resilience under external shocks.

With respect to FRM, the ontological view of scale suggests that (a) tailored measures must be considered for individual spatial domains, taking into consideration that FRM measures at one scale and for certain flood types may exhibit different and perhaps contradictory effects at another scale and for different types of floods. Such spatiotemporal changes in the effects of FRM measures are, to a large degree, effects of duality in view of land as a social (legal) institution and as a natural entity, as will be discussed in the last section of the chapter. In reference to the variability of scales, (b) attention must be paid to the transferability of expert knowledge from one scale to another. Finally, (c) any process and agent (landowner, municipal authorities, river basin authorities, NGOs) cannot be implicitly considered local because of its apparent size or 'root location'—simply, the scale of actions and impacts of the locally-rooted natural and social processes may display distinct variation, given inherent linkages to higher spatial scales. For example, local polders—if well-designed—may have impacts on a catchment exceeding their own spatial extent; the blue-green infrastructure—if sufficiently connected—may influence microclimatic patterns (Louda and Macháč—Part III, Chapter 1), and even one leader may influence the decision-making in the whole catchment depending on his or her communication skills, networks and his or her ability to mobilize available lay knowledge.

Upscaling or Downscaling in Flood Risk Management?

The discussion above opens a crucial question: what is the right and available way to shift our knowledge and practice between scales? This means, how should the processes of upscaling and downscaling in FRM be addressed? Before trying to find an answer, however, our and any practitioner's thoughts should finally determine if and why it is important to downscale and upscale the FRM measures, including the NBS. While downscaling may be easily supported with a necessity of involving local stakeholders, the intriguing question is if and why bigger is considered better in FRM (note that bigger does not stand for "uniform" here and in the following text). There are several arguments (Buijse et al. 2002; Adger et al. 2005; Zevenbergen et al. 2008) that justify the environmental rationale for extensive FRM measures, taking into account the complementarity of their direct hydrologic effects and indirect effects on ecosystem services (Schanze 2017).

Generally, both ecological and economical studies suggest that open systems with existing multiple links are more resilient and have a higher performance in a long-term perspective although they may suffer from short-term external disturbances. For example, although gross domestic product decreased and the unemployment rate rose in many countries during and after the global economic crises in 2007–2008, the speed of recovery and current economic performance of these countries is still higher than that experienced by semi-open/closed totalitarian economies. Similarly, although depending on scales, the effect of external disturbances seems to be more visible in regions of industrial specialization than those with economic diversity (Ženka et al. 2015). In an ecological domain, it was argued that broad-scale programmes may be more efficient because they make use of environmental variation in order to compensate for local or temporary risks (e.g., water scarcity, accelerated sedimentation during floods, land degradation; e.g., Cumming et al. 2006).

Various methods have been developed to understand the uncertainties resulting from transfer of the site-specific or experimental laboratory research results to broader scales or to design a multi-scale hierarchical research that would clarify the variances of bioecological, biophysical and social processes across scales. Among these methods, two approaches are the most frequent (cf., Burt 2003). The statistical approach uses a generalization based on filtering or geostatistical analyses in order to remove small-scale noise and to emphasize the general statistical patterns (trends and fluctuations). This approach may be partly useful when attempting to upscale patterns of individual features and processes, such as flood frequency analyses or changing social vulnerability to floods from time-series. It would be insufficient, however, for the study of complex causal links within the socio-ecological systems, because the nature of such links may exhibit variances across scales (see above). In fact, apparently small-scale noise or a dependent variable at a local scale may represent independent or causal variables at other spatial and time scales.

Another approach to the multi-scale hierarchical research and to upscaling its results for practical issues is a nested experiment design. This approach resolves two troubles faced by statistical approaches: first, it establishes an a priori hierarchical (quasi-continual) design, thus not inferring the nature of processes at one scale from data gained at another scale; second, it overcomes the limits of financial sources and time to perform detailed empirical sampling over the vast areas. The nested approach stems from the assumption that by looking at various scales (study plots/areas) in one area that are close enough in order to study the links among each pair of neighbouring scales, we will finally be able to fill the gap between the two outer (smallest and largest) scales (Costanza and Folke 1996, p. 24).

Scale Mismatch in Socio-ecological Systems

Deriving from different views on scale and continuing to variations in physical and social processes across spatiotemporal scales, we have arrived at a critical step that limits the effort for implementation of NBS in FRM. In particular, if we accept the assumption that broad-scale NBS are beneficial for FRM and that approaches for upscaling our experience (with certain limits) to broader scales are available, we may pose a fundamental question: why do we still encounter such difficulties in the implementation of broad-scale NBS in FRM and why are well-documented good practice cases still so rare?

Basically, the answer lies within the duality in the perception of ecological and social systems. While it is easy to understand that territorial integrity (or at least spatial density) of NBS for water retention will do a better job than site-specific and separated measures, the national and regional policies, land tenure models as well as property laws have resulted in extreme land fragmentation. As Freyfogle has stated, "one has to do with the mismatch between the way private land is portrayed in law and culture and the way it exists in real world of nature" (Freyfogle 2003, p. 7). In addition, in many political and tenure systems, such duality in the view of land would not only relate to private land but rather to the continuity between the public and the private. The fragmentation resulting from both the ownership and administrative division, in turn, limits the viability of broad-scale NBS if not all integrated FRM measures. By transforming land into a resource (Hanna and Jentoft 1996) people imply its differential values and set respective property rights to assure expected profits for stakeholders and owners (Freyfogle 2003, p. 242), thus disconnecting the social and ecological systems. Such disconnection builds a sharp divide in ways of argumentation for the right to access and benefit from shared environmental values (ecological services) provided by nature.

In an FRM context, such duality is illustrated when the suggested measures are upscaled based on empirical data that was gained at more detailed scales and in two domains with different scale dependencies. Some failures in FRM should then be considered to result from a mismatch between the scale of management and the scale of environmental variations (Gunderson and Holling 2002; Young 2002; Cumming et al. 2006). The mismatch is obvious at lowest administrative levels; for instance, where municipalities have gained fundamental (even though not sole) competencies in the past few decades (cf., the EU Floods Directive 2007/60/EC). Despite the arguments for the decentralization of FRM, it is now increasingly difficult to coordinate the ecologically relevant measures in catchments, which are fragmented across a number of administrative units (Moss and Newig 2010; Huesker and Moss 2015; Slavíková et al. 2019) and diversified into various FRM strategies and practices (Gilissen et al. 2016). The effects of land and institutional fragmentation will certainly differ for various flood types. It can be assumed that the administrative fragmentation will strongly affect the effectiveness of the FRM programmes aimed at riverine flood risk reduction. On the other hand, efforts aimed at flash flood risk reduction or storm surges in small semi-natural catchments or specifically in urban areas will increasingly face the ownership fragmentation, whereas the administrative fragmentation may be limited.

Accordingly with Cumming et al. (2006), we may postulate that scale mismatches may derive from the social domain (e.g., tenure systems, policies), the ecological domain (changing nature of processes across scales, see Sect. 2), but also from coupled social-ecological processes. In the social domain, what individuals expect and how do they decide in terms of their own property and their community, and what they expect from the government (NIMBY syndrome is only one example illustrating contradictory behavioural responses; see, e.g., Rand and Hoen 2017) is generally contradictory. Accordingly, flood risk reduction behaviour and decision-making will exhibit differences at the individual, municipal, catchment or national scale respectively. Also in the ecological domain may flood impacts imply significant variations (trade-offs) across scales. Flash floods may bring about necessary episodic local disturbances for the sake of the renewal of forest ecosystems upstream, but downstream deposits of high volumes of material may arrive, for instance. The mismatches resulting from a complex grid of spatiotemporal causes and of social and/or ecological domains will appear as primarily spatial, temporal or functional, but in all cases they will denote a situation where hierarchies of management and ecological processes are not aligned and do not allow for effective and efficient planning of FRM measures.

Concluding Remarks on Realignment of Scale in FRM

Considering the above discussed scale mismatches, FRM may be successful if it is conceived in an environment where the scale is capable of effective implementation within and respecting the existing land tenure system. For this reason, several researchers and planners have tried to find a viable common scale that would allow for social and political acceptance of FRM (e.g., a catchment scale or a municipal scale, the EU Floods Directive 2007/60/EC; Hartmann and Juepner 2014), and to discuss options for incremental and complex planning processes (cf., Lindblom 1959).

Realignment of scales to improve the FRM would be difficult for many reasons, however. First, determining the source of scale mismatch can be tricky when the physical processes are not well understood, competencies and agendas among institutions are not clearly defined, or possibilities to transform inherited policy practices are limited due to multiple path-dependencies in social, political and economic systems. In this context, the implementation of the EU Water Framework Directive (2000/60/EC) and the EU Floods Directive (2007/60/EC) represents the unique experiment of social and ecological re-scaling with (so far) uncertain effects on the resource quality and abundance (see Huesker and Moss 2015 or Jager et al. 2016 for detailed information).

For the above-listed reasons, it is of fundamental importance to reassess the options for upscaling NBS in FRM and reconsider them in terms of possible scale mismatches. If land uses and land ownership and management structures are not properly addressed, the effort for nature-based solution support in FRM will remain inefficient. Therefore, it seems that these policy objectives may only be reached if we broaden the available evidence about the hydro-ecological effects of NBS in FRM at various scales, while addressing acceptability and willingness to adopt these solutions at comparative institutional scales and under different legal regimes. Clearly, our ambition must be to explore the methodological options for an interdisciplinary approach within the individual studies in addition to fragmented hydro-ecological research, social inquiries or legal studies.

Acknowledgements We thank the Operational Programme Research, Development and Education of the Czech Republic for financing the project Smart City—Smart Region—Smart Community (grant number: CZ.02.1.01/0.0/0.0/17_048/0007435) that led to the present book chapter.

Open access of this chapter is funded by COST Action No. CA16209 Natural flood retention on private land, LAND4FLOOD (www.land4flood.eu), supported by COST (European Cooperation in Science and Technology).

References

Adger WN, Arnella NW, Tompkins EL (2005) Successful adaptation to climate change across scales. Glob Environ Change 15(2):77–86

Alfonso I (ed) (2007) The rural history of medieval European societies—trends and perspectives. Brepols Publishers, Turnhout

Brenner N (2001) The limits to scale? Methodological reflections on scalar structuration. Prog Hum Geogr 15:525–548

Buijse AD, Coops H, Staras M, Jans LH, Van Geest GJ, Grit RE, Ibelings BW, Oosterberg W, Roozen FCJM (2002) Restoration strategies for river floodplains along large lowland rivers in Europe. Restor Ecol 47(4):889–907

Burt T (2003) Scale: resolution, analysis and synthesis in physical geography. In: Holloway SL, Rice SP, Valentine G (eds) Key concepts in geography. SAGE Publications, London/Thousand Oaks, pp 209–228

Costanza R, Folke C (1996) The structure and function of ecological systems in relation to property-right regimes. In: Hanna S, Folke C, Maler KG (eds) Rights to nature. Island Press, Washington, DC, pp 13–34

Cumming GS, Cumming DHM, Redman CL (2006) Scale mismatches in social-ecological systems: causes, consequences, and solutions. Ecol Soc 11(1):14

de Moel H, Jongman B, Kreibich H, Merz B, Penning-Rowsell E, Ward PJ (2015) Flood risk assessments at different spatial scales. Mitig Adapt Strateg Clim Change 20(6):865–890

EU Floods Directive 2007/60/EC (2007) Directive on the assessment and management of flood risks. European Community

Ernst D, Kim L (2002) Global production networks, knowledge diffusion, and local capability formation. Res Policy 31(8–9):1417–1429

Freyfogle ET (2003) The land we share: private property and the common good. Island Press, Washington, DC

Giddens A (1984) The constitution of society. Outline of the theory of structuration. Polity, Cambridge

Gilissen HK, Alexander M, Beyers J-C, Chmielewski P, Matczak P, Schellenberger T, Suykens C (2016) Bridges over troubled waters: an interdisciplinary framework for evaluating the interconnectedness within fragmented domestic flood risk management systems. J Water Law 25(1):12–26

Green WA (1995) Periodizing world history. Hist Theory 34(2):99–111

Gregory D, Johnston R, Pratt G, Watts MJ, Whatmore S (eds) (2009) The dictionary of human geography. Wiley-Blackwell, Oxford

Gunderson L, Holling CS (eds) (2002) Panarchy: understanding transformations in human and natural systems. Island Press, Washington, DC

Gurevic AJ (1985) Categories of medieval culture. Routledge and Keegan Paul, London

Gusyev M, Gädeke A, Cullmann J, Magome J, Sugiura A, Sawano H, Takeuchi K (2016) Connecting global- and local-scale flood risk assessment: a case of the Rhine River basin flood hazard. J Flood Risk Manag 9(4):343–354

Hägerstrand T (1970) What about people in regional science? Pap Reg Sci Assoc 24(1):6–21

Hanna S, Jentoft S (1996) Human use of the natural environment: an overview of social and economic dimensions. In: Hanna S, Folke C, Maler KG (eds) Rights to nature. Island Press, Washington, DC, pp 35–55

Hartmann T, Juepner R (2014) The flood risk management plan: an essential step towards the institutionalization of a paradigm shift. Int J Water Gov 2(1):107–118
Harvey D (1989) The condition of postmodernity. Blackwell Publishers, Hoboken, NJ, USA
Huesker F, Moss T (2015) The politics of multi-scalar action in river basin management: implementing the EU Water Framework Directive (WFD). Land Use Policy 42:38–47
Herod A (2003) Scale: the local and the global. In: Holloway SL, Rice SP, Valentine G (eds) Key concepts in geography. SAGE Publications, London/Thousand Oaks, pp 229–246
Jager NW, Challies E et al (2016) Transforming European water governance? Participation and river basin management under the EU Water Framework Directive in 13 member states. Water 8(4):156
Lindblom C (1959) The science of "muddling through". Public Adm Rev 19(2):79–88
Lynch K (1960) The image of the city. MIT Press, Cambridge, MA
Marston SA, Jones JP III, Woodward W (2005) Human geography without scale. Trans Inst Br Geogr 30:416–432
Mayhew S (2015) Oxford dictionary of geography. Oxford University Press, Oxford
Moss T, Newig J (2010) Multilevel water governance and problems of scale: setting the stage for a broader debate. Environ Manag 46:1–6
Muhar S, Januschke K, Kail J, Poppe M, Schmutz S, Hering D, Buijse AD (2016) Evaluating good-practice cases for river restoration across Europe: context, methodological framework, selected results and recommendations. Hydrobiologia 769(1):3–19
Nesshöver C, Assmuth T, Irvine KN, Rusch GM, Waylen KA, Delbaere B, Haase D, Jones-Walters L, Keune H, Kovacs E, Krauze K, Külvik M, Rey F, Van Dijk J, Vistad OI, Wilkinson ME, Wittmer H (2017) The science, policy and practice of nature-based solutions: an interdisciplinary perspective. Sci Total Environ 579:1215–1227
Office of Environment and Heritage (OEH) (2017) A proposed new legal framework: aboriginal cultural heritage in NSW. OEH, Sydney
Rand J, Hoen B (2017) Thirty years of North American wind energy acceptance research: what have we learned? Energy Res Soc Sci 29:135–148
Schanze J (2017) Nature-based solutions in flood risk management—buzzword or innovation? J Flood Risk Manag 10:281–282
Schumm SA (1994) Erroneous perceptions of fluvial hazards. Geomorphology 10:129–138
Schumm SA, Lichty RW (1965) Time, space, and causality in geomorphology. Am J Sci 263(2):110–119
Sheehan J (2014) The effects of sea level rise and increased storm events: an editorial introduction. Geogr Res Forum 34:1–8
Slavíková L, Raška P, Kopáček M (2019) Mayors and "their" land: revealing approaches to flood risk management in small municipalities. J Flood Risk Manag e12474 (in press)
Termeer CJAM, Dewulf A, van Lieshout M (2010) Disentangling scale approaches in governance research: comparing monocentric, multilevel, and adaptive governance. Ecol Soc 15(4)
Wilbanks T, Kates RW (1999) Global change in local places. How scale matters. Clim Change 43:601–628
Young OR (2002) The institutional dimensions of environmental change: fit, interplay, and scale (global environmental accords: strategies for sustainability). MIT Press, Cumberland
Ženka J, Novotný J, Slach O, Květoň V (2015) Industrial specialization and economic performance: a case of Czech microregions. Nor J Geogr 69(2):67–79
Zevenbergen C, Veerbeek W, Gersoniu B, Van Herk S (2008) Challenges in urban flood management: travelling across spatial and temporal scales. J Flood Risk Manag 1(2):81–88

Pavel Raška (*1982) graduated from JEP University (history and geography) and Masaryk University (physical geography). He is currently Associate Professor and Head at the Department of Geography, Faculty of Science, J. E. Purkyně University in Ústí nad Labem. His main research interests include flood hazards and risk reduction, in which he explores interdisciplinary perspectives on institutional barriers to FRM and planning.

Lenka Slavíková graduated from University of Economics, Prague (public economics and policy). Currently, she serves as the Associate Professor at the Faculty of Social and Economic Studies J. E. Purkyně University in Ustí nad Labem. Her long-term interest is in water and biodiversity governance with the focus on Central and Eastern European Countries. She investigates flood risk perception of different actors and financial instruments for flood recovery and mitigation.

John Sheehan currently Adjunct Professor in the Faculty of Design Architecture and Building. He is also Adjunct Professor in the Faculty of Society and Design at Bond University, Gold Coast, Queensland. In 2017, John was appointed Guest Professor with the Institute for Economic and Environmental Policy (IEEP) in the Faculty of Social and Economic Studies at J. E. Purkyně University in Usti nad Labem, Czech Republic. In 2003 he was appointed the independent Chair of the Project Advisory Committee Water Property Titles Program, Land and Water Australia, which was a research project funded by the Commonwealth Government to establish a water titling system for Australia.

Part II
Water Retention in the Polish Forests

Chapter 3
Reversing the Current: Small Scale Retention Programs in Polish Forests

Piotr Matczak, Viktória Takács and Marek Goździk

In this chapter, we present the small retention programs that have been undertaken in the Polish forests. The context of the programs is outlined with an emphasis on property rights, actors engaged, nature conservation and flood management aspects, and finally on the issues of up-scaling. After 1997 several small retention programs were introduced, reversing the earlier practice of drainage. The programs in the State Forests National Forest Holding consist in building a diverse range of small-scale, multi-functional retention facilities. The FRM aspect is more important in the mountain areas, while reducing forest fire risk is more important in the lowlands. Location of facilities was, in general, not problematic, as they were mostly built in the forests managed by the State Forests National Forest Holding. However, in particular locations, conflicts of interest or collisions between functions (wood production, nature conservation, etc.) occurred. Although the State Forests National Forest Holding manages the small retention programs, other stakeholders are engaged in planning small retentions also: the environmental protection administration, local governments, farmers, fishpond owners, etc. The program is based on a clear hierarchical managerial structure of the State Forests National Forest Holding, which makes the up-scaling aspect of the program straightforward.

P. Matczak (✉)
Institute of Sociology, Adam Mickiewicz University in Poznań, Poznan, Poland
e-mail: matczak@amu.edu.pl

V. Takács
Institute of Zoology, Poznań University of Life Sciences, Poznan, Poland
e-mail: takacsviki@o2.pl

M. Goździk
Coordination Centre for Environmental Projects, The State Forests National Forest Holding, Warsaw, Poland
e-mail: marek.gozdzik@ckps.pl

T. Hartmann et al. (eds.), *Nature-Based Flood Risk Management on Private Land*,
https://doi.org/10.1007/978-3-030-23842-1_3

Introduction: Water Management in Poland in the Retention Context

Water resources per capita in Poland are among the lowest in the EU. They are also unevenly spatially distributed: central Poland can be characterized as water deficient at a relatively lower level of precipitation than other regions (Mioduszewski and Pierzgalski 2009). Occasional droughts cause serious losses in agriculture.

At the same time, floods present the biggest natural hazard in Poland. The fluvial type of floods, caused by storms and sudden high precipitation, dominate in the country. It is estimated that in the post-war period (1946–2010), 600 floods have occurred, of which fifteen floods were classified as disastrous on at least a regional scale. Particularly, the floods in 1997 (on the Odra River) and in 2010 (on the Vistula and the Odra Rivers) caused serious material losses and dozens of casualties (Kundzewicz et al. 2012).

Agriculture intensification and homogenisation of landscape, construction of drainage systems as well as urban development have all resulted in sealing ground surfaces and have thus intensified flood risks significantly. The natural water retention capacity of catchments has decreased while the runoff paths have been streamlined causing a rapid runoff of precipitation and melting snow into rivers (Mioduszewski 2014).

Consequently, retention is badly needed in Polish water management in order to mitigate floods (Fournier et al. 2016). The very concept of retention as an approach in water resources and FRM stems from a conviction that occasional excess of water can be stored; this measure decreases the flood risk and secures water supply during water shortage. Retention is thus instrumental in FRM, agriculture production and also has other functions: water storage in case of wild fires, biodiversity protection, fish cultivation, positive influence on biodiversity and on landscape characteristics, which further results in more attractive recreational areas.

In this chapter, water retention in forested areas is presented—a new approach in flood management and water resources management that has been developed and applied for the last 20 years in Poland. The programs have relied on the concept of small retention. It is a term referring to various means and techniques aiming to limit the water runoff after heavy rains or snow melting. The main idea is to improve hydrographic conditions in a catchment by increasing the time and the track of water circulation (Mioduszewski 2014; Kowalczak 2002). Small retention facilities slow down water outflow from natural and artificial running waters, store waters in small reservoirs and terrain depressions and increase the retention of water in soils and aquifers (Kowalewski 2008).

”Small retention” is a term seldom used in other countries. In Poland, small retention is used in contrast to large multifunctional reservoirs, serving as drinking water intakes, power production facility, flood control etc. (Kowalewski 2008). This term has a meaning not far from the concepts of “natural retention” (Natural Water Retention Measures) (Koseoglu and Moran 2014) and “water harvesting”. Yet, the term of small retention has a wider scope as it denotes both technical measures and use of natural formations for retention purposes.

In the following sections changes in the water management are presented, followed by the description of the small retention program in Polish forests. It is discussed in terms of implementation, measures, costs, function, property rights issues and up-scaling.

The Small Retention Programs in Polish Forests

The current efforts to retain water stem from the legacy of the communist period from 1945 to 1990. In this time period, productivity was the main priority, resulting in significant development of drainage in agricultural areas (Mioduszewski 2014). The development of drainage systems contributed to the transformation of the water regime. The majority of small watercourses were regulated, and a number of dense networks of drainage ditches were constructed. As a result, in large areas the groundwater table has been excessively lowered (Mioduszewski 2014).

Polish forests were also drained, to increase timber productivity and grow forests on the wetlands. This approach was in accordance with the socialist economy features, including planned, industrialized economy development. As a result of this program, more than 850 thousand hectares of forested area were drained with further plans to continue. Drainage in forests led to the degradation of forest wetlands. The water table decreased, and many small wet areas and peatbogs disappeared (Miler 2015).

After 1990, with the collapse of the communist system, the drainage programme lost its importance. This was in line with the wider, radical change of the economic, political and administrative order in the country. From the beginning of the 1990s, new drainage facilities were built only occasionally, and the existing ones were not maintained and mostly became abandoned. New water and FRM policies were gradually built.

The Small Retention Programs in Polish Forests

Forests cover 9.2 million hectares, which is 29.5% of the territory of Poland; 82% of the forests are public. The State Forests (SF) National Forest Holding is a specific organisational structure established to manage the public forests, and it also plays a supervisory role for all forests. The dominant position of the public forestry sector—and especially the SF National Forest Holding, which is responsible for the management of most forests in the possession of the State Treasury—is the distinctive feature of the forest policy in Poland. After the collapse of the communist regime in 1990, the restitution of formerly nationalised forest estates has not taken place in Poland, unlike other Central European countries. It is due to the high degree of fragmentation of private forests, where the average area of a single forest holding is 1.3 ha (Zając 2004). The resulting position of the SF National Forest Holding enables the agency to manage the majority of the Polish forests thus to implement a coherent National Forest Policy.

After 1990, the forest policy objectives have shifted. In a document regulating planning and forest management, the Forest Silva-culture Principles, sustainable elements of forestry were first mentioned. While timber production had previously been the priority, additional focus on sustainability principles and economically balanced the models gradually developed thereafter. The national forest policy concerning forests includes sustainable management and the maintenance of multi-functionality of forests. Small retention is a component of this shift, aligned with the general policy. In 1995, agreements between the Ministry of Agriculture and the Ministry of the Environment concerning cooperation about small retention were developed. Eventually, in 1997, the Director of the SF National Forest Holding initiated the small retention programme for the period of 1997–2007. The program was well received and it was decided to continue it. The second period was executed in the years 2008–2014, and the third period was planned for 2015–2022.

In the first period, 3340 retention objects were constructed: 727 small reservoirs were built, 4 lakes were dammed, and 1551 smaller ponds were created. Furthermore, over 800 objects were connected to basic and extended reclamation networks. Damming lakes and building reservoirs gave the highest result in retention capacity—over 50 million cubic metres both, while the rest of devices had far less capacity (Kowalewski 2008).

In 2006, the second period of the small retention program in the Polish forests was launched, consisting of two parts dedicated to the lowland and upland forests. Although the flood protection aspect was present in both parts, the emphasis was differently placed. In lowland forests, increasing retention capacities was intended mainly to prevent droughts and to fight wildfires. The program for the maintenance areas was designed to counteract the effects of rainwater runoff from the mountains, increase water retention capacity in watersheds, maintain torrents-related infrastructure in good condition, and reduce erosion (Pierzgalski et al. 2017).

Implementation of the Small Retention Program in the Polish Forests

The small retention methods can be roughly divided into two categories: natural and technical. The natural forms include any sort of land use change such as forestation areas alongside rivers, slope shaping, protection planting, revitalization and preservation of swamps, moors and marshes, soil structure improvement wetland restoration, re-meandering and others. The technical forms include building ponds, oxbow lakes, moats, clay pits, fire water reservoirs, damming on rivers and lakes, establishing the river polders, building buffer ponds in headwater catchments, and installing technical devices such as weirs, gates, and barrages.

Implementation of small retention projects in forests has several advantages, such as the following: ensuring forest growth by raising water table; decreasing forest fire risk; increasing groundwater resources, runoff water purification; biodiversity conservation; providing waters intakes for irrigation, re-naturalization of wetland habitats and regulation of rivers, development of fishing resources (Mioduszewski 2014). Nevertheless, small retention can also have adverse effects related mainly to technical measures. In particular, building new reservoirs, besides restoring existing ones, can have negative effects on biodiversity; this matter will be discussed later.

Within the small retention programs, several measures were planned. Nevertheless, their small scale was the core of the programs. Small water storage facilities were planned at a maximum capacity of five million cubic metres, although in practice they have not exceeded one million cubic metres. Small retention ponds were intended mainly to improve water balances in the forest ecosystems (Pierzgalski et al. 2017).

The programs relied on several conditions when building the small retention facilities, including the following: (a) forests and buffer woods planting should reduce surface runoff; (b) the existing reclamation, irrigation, and water-lifting facilities should be maintained, in particularly those hampering water outflow; (c) local communities and regional boards of water management should be consulted when planning facilities; (d) facilities should be located in places facilitating the recharge of aquifers. These conditions were defined in the program document and in the good practice books prepared by the SF National Forest Holding units (CKPŚ 2008; Goździk et al. 2009; CKPŚ 2016).

About seven thousand measures and hydraulic devices were built within the programs in both up and lowland regions (see Table 3.1).

For the programming period 2014–2020, the upgrade and the new financial plan for the small retention program was designed. Similarly to the previous periods, the program is divided into mountain and lowland areas. However, climate change represents an additional rationale for action in this phase. The projects are called "Counteracting water erosion in mountain areas" and "The comprehensive project of adaptation of forests and forestry to climate change—small retention and counteracting water erosion in the lowland areas" (CKPŚ 2016).

The main goal for the next period is to strengthen the resilience to threats related to climate change in both low- and upland forest ecosystems. The activities are

Table 3.1 Small retention objects, retained water capacity and costs in upland and lowland forests within the programming period 2007–2013

	No. of objects	Retained water in million of m^3	Total cost
Small retention on lowlands	3644	42.8	44 million euro
Small retention on highlands	3553	1.5	43 million euro

SF National Forest Holding: http://www.ckps.lasy.gov.pl

aimed at preventing or minimizing the negative effects of natural disasters such as the following: floods and inundations, droughts and wild forest fires via the development and maintaining the existing small retention systems, increasing the amount of stored water and counteracting excessive water erosion. Besides new retention objects, the program also hopes to reconstruct valuable natural ecosystems and biodiversity conservation and maintain the existing constructions. Monitoring is also an important element of the program, partially as the continuation of monitoring performed within the framework of the small retention program implemented within the previous periods.

Measures Applied in the Small Retention Programs

The small retention program has several preventative functions in case of floods, drought, wild fires, erosion, and other disasters, and for the conservation of biodiversity. Within the programs, several technical and non-technical measures were implemented, such as construction of close to nature ponds, lateral reservoirs, forestation revitalization of existing and dried pit areas (CKPŚ 2016).

Particular measures usually have functions that depend upon a specific situation in a location. Application of measures depends on local circumstances but is also driven by the building principles and the good practice books (CKPŚ 2008; Goździk et al. 2009; CKPŚ 2016) that were prepared within the program. For instance, building with the use of natural materials, minimizing the use of concrete, is advised as well as taking into account natural processes during implementation.

Costs and Funding

In the years 1997–2007 costs of the retention programs in Poland were covered mostly by public sources. Facilities were co-financed by budgets of provincial governments, the Provincial Funds for Environment Protection and Water Management, the National Fund for Environmental Protection and Water Management, local governments, the Fund for the Protection of Agricultural Lands (FPAL), and the Agency for Restructuring and Modernization of Agriculture (ARMA). About 25% of costs were covered by private investments, mostly owners of private fish ponds, small hydro-energy production stations, fishing associations, mining companies, etc. Projects were financed on a competitive basis.

The largest sums spent came out of the budgets of provincial governments and the Provincial Funds for Environment Protection and Water Management. On average, the program's mean annual expenditure equalled EUR 12 million (Kowalewski 2008; Mioduszewski 2014).

Contrary to the first programming period, in the years 2007–2013, the program was financed mainly (85%) by the EU Cohesion Fund, within the Operational Program Infrastructure and Environment, Action 3.1. In the end, the SF National Forest Holding financed approximately 30% of the program's costs. Funding was divided between lowlands and uplands. In both cases, it equalled about 44 million euro.

Funding for the programming period 2014–2020 is estimated to approximate EUR 55 million in lowland and EUR 49 million in upland areas.

Up to the year 2006 the most effective measure, for the whole country and among the various programmes, was lake damming, at a cost of 0.015–0.13 euro per cubic metre. Lake retention requires relatively low inputs in terms of infrastructural investments. When the lake retention options were exhausted, other measures appeared less cost effective. They entailed higher costs per cubic metre of retained water. For instance, the cost of obtaining one cubic metre of retained water in the case of fishponds varied between EUR 0.5 and 1.4 per one cubic metre; the costs of the construction of artificial reservoirs appeared even higher and varied, from EUR 0.86 to 4.7 per one cubic metre (Kowalewski 2008; Tyszka 2009). On average, for the phase of 2007–2013, cost of retaining of one cubic metre equalled EUR 1.5 for the lowlands and EUR 10.4 for the mountain areas.

In terms of costs for various sizes of reservoirs, estimates suggest that small retention is about ten times cheaper than larger reservoirs. The cost of storage of one cubic metre of water in small retention facilities ranges from EUR 0.5 to 1.2, while large retention reservoirs range from EUR 3.6 to 9.5 per cubic metre (Liberacki et al. 2016; Miler 2015).

Actors, Stakeholders and the Property Rights Aspect of the Programs in Poland

Besides the units of the SF National Forest Holding, several administrative agencies were involved in retention programs in the country such as the water management bodies (The Regional Directorate of Water Management), flood risk managers (Institute of Meteorology and Water Management; Regional Drainage and Water Facilities Boards), the Provincial Funds for Environment Protection and Water Management, National Parks, Landscape Parks, Promotional Forest Complexes, municipalities, the county and the provincial administrations, and the special interests groups: naturalists; farmers, fish pond owners etc. Moreover, the small retention project had to be included in several long-term plans, in particular, the forest management plans, spatial development plans of municipalities, and management plans of the National Parks (Zając 2004).

Concerning the retention programs, cooperation on the local level with several environmental NGOs helped to plan and realise the programs in the Polish State Forests. Experts and NGOs helped in working out methods of locating and building small retention facilities (CKPŚ 2009; Jelonek et al. 2008; JASPERS 2009); NGOs were involved both during the planning and implementation phases. They provided advice in identification of new reservoirs locations and carrying out environmental procedures.

The property rights issue appears relatively non-problematic in the programs. The majority of facilities were located in the forests managed by the SF National Forest Holding. From the perspective of the Holding, as the owner and the manager of forests, the small retention programs were not the core issue for the organisation. However, the programmes were in accordance with several of the Holding's objectives, such as nature conservation, forest management, etc. The decision to initiate the programs made it possible to use the vast forest areas for another purpose. Although it was seemingly not a deliberate decision from a FRM perspective (as the program aimed mostly at fire protection and drought management), the fact that facilities were built on the land of one owner was a fortunate condition. In cases where other stakeholders were involved, usually conflicts of interest occurred. Nevertheless, protection of the third parties' interests was mentioned in the good practice advice books prepared by the SF National Forest Holding units (CKPŚ 2008; Goździk et al. 2009; CKPŚ 2016).

The Nature Conservation Aspect

In practice, retention sites are often planned on already existing small wet areas, usually with a high natural value. The reason for this is that from the hydrological point of view the most suitable places for retention sites are existing small inter-forest wetlands. These sites are also appropriate from economic considerations as they usually hold a limited potential for production. They are often situated within Natura 2000 sites since 25% of lowland forests and over 90% of highland forests are part of Natura 2000 network. Therefore potential impacts on biodiversity are important issues.

In general, wetlands, swamps and small ponds in the forest are very important for biodiversity (Baldwin 2005). Wet sites in the forests directly increase both species and habitat diversity. This has a special importance in the case of forest monocultures where wet sites are one of the main sources of diversity within the monotonous landscape (Whitaker and Montevecchi 1997). Wet sites are habitats hosting a diverse wildlife. All taxonomic groups have representatives inhabiting inter-forest wet areas that are essential, especially for amphibians and migrating waterfowl (Baldwin et al. 2006; Di Mauro and Hunter 2002).

Additionally, wildlife benefits indirectly from the vicinity of wetlands and increasing forest moisture (Mioduszewski 2014) by augmenting forest growth and vegetation density. A survey of birds (other than waterfowl) showed that the vicinity of wet in forest increased both bird diversity and density in production forests. Apart from direct and indirect influences on biodiversity, wet sites in forests help to mitigate the negative impacts of forest management (Hanowski et al. 2006).

On the other hand, the creation of small retentions does not always comply with principles of biodiversity conservation. During the construction of retention projects, construction can destroy some natural ecosystems. Retention projects can also influence biodiversity indirectly. As this process always involves the creation of new habitats and changes in the water table, cleaning existing channels can lead to less predictable effects on biota (Wegner 1999; CKPŚ 2009): for example, destroying peat habitats by flood or changing the water regime can change water habitats for fish and water plants. These impacts are often difficult to detect and predict.

An additional potential risk is connected to changing the network of wet areas at a landscape scale. Newly filled ditches can allow for the migration of alien species such as the American mink (Neovison vison) along the newly opened ditches (Ahlers et al. 2016). The American mink is an invasive alien predator in Poland that mainly escapes from fur farms and endangers waterfowl and other birds, especially in areas distant from human settlements (Nordström et al. 2002; Brzeziński et al. 2012).

A similar phenomenon is the probable proliferation of beavers (Castor fiber) along small retention sites. The beaver's population, though previously almost extinct in Poland, has increased over the last 30 years to the extent that all possible sites—running water and channels—will be occupied by them. Beavers can increase overall forest biodiversity and water quality (Puttock et al. 2017). However, their growing population also has some negative consequences for forest management and planning.

Hydrological Aspects and Monitoring the Efficiency of the Small Retention Programs

The aim of small retention programs is to help in local drought and flood protection and lower the risk of wild fires. The small retention program was not clearly focused on flood management but rather aimed to establish a complex and ecologically sound water resource management in forests. In the lowland areas, small retentions decrease flood risk and mitigate droughts. In the mountain areas, water retention aims mainly at decreasing flood risks.

A forest is a natural storage reservoir and can be treated as a space for water retention. Water storage in forests can mitigate floods, as it flattens the flood wave. The type of trees, their age, height, compactness, undergrowth, litter, etc. have an impact on the success of the mitigation. Fresh, dry and mixed forests as well as moist mixed forests have the highest water storage capacity. They can store 28–30% of rainfall. Riparian and alder forests have much lower (10–13% of rainfall) capacity (Mioduszewski and Pierzgalski 2009).

It is not fully determined how to measure of the efficiency of precautionary water retention measures in forests as it depends on many hydrological and geological factors and also on the scale of monitoring (Schüler 2006). Afforestation in a catchment is considered as an element contributing to water circulation. Poor permeable soils in the catchment area, with variations of land elevation, leads to a high degree of surface runoffs. The small retention reservoirs may further limit the flood wave and can therefore serve as an element of the flood risk reduction system beyond their purpose as land use planning measures. To perform the flood reduction function, small tanks must be equipped with valves enabling water retention only during the peak flow period (Mioduszewski and Pierzgalski 2009).

The quantitative assessment of small retention is usually not as evident (Miler 2015). A number of publications referring to the SWAT (Soil and Water Assessment Tool) models were developed to assess the impact of small retention measures on flood protection and mitigating the effects of drought. These models analyse elements of retention process, like the impact of wetland restoration on the flood wave or the impact of cultivation types on river flow. At the same time, the diversity of factors shaping the outflow process makes it impossible to draw universal conclusions about the impact of small water retention. Small retention processes are specific for a given catchment and climate, and scale of events (Mioduszewski and Okruszko 2016).

Concerning the small retention programs launched in Poland from 1997, a network of retention facilities covering the Polish forests has been planned to increase retention possibilities and counteract floods and droughts in forest ecosystems in lowland areas and the mountains (Mioduszewski and Pierzgalski 2009). Numerical modelling shows that the program is instrumental in providing water supply for plants and ecosystems during drought; however, they further prove empirically that the program's impact on flood protection is still a challenge (Mioduszewski and Okruszko 2016). Nevertheless, a number of studies have been published on monitoring the influences of retention in Polish forests. According to these studies, the smaller the pond and the smaller the value of the current water body retention, the bigger will be the relative increase of groundwater retention in the areas adjacent to a pond in relation to the increase of the water level in said pond (Juszczak et al. 2007).

An analysis of a small water retention resulting from installing weirs in the watercourse in small forest catchments of the Krajeńskie Lake District showed that the average value of the time constant for flood waves increased by about 50% following the construction of the reservoir (Miler 2015).

Small retention is limited in terms of achieving flood protection as a natural process is largely uncontrollable and difficult or impossible to regulate. A forest has a potential threshold retention capacity; thus the impact of a forest on flood flows is also limited to a certain amount of precipitation (Mioduszewski and Pierzgalski 2009).

Synergies and Tensions Between Functions

Tensions and collisions between the small retention program activities and biodiversity conservation, as mentioned above, are related to the fact that a large number of forests are part of the Natura 2000 network. In accordance with the Natura 2000 conservation principles, all investments need to be screened in terms of their environmental impact. As a result, construction of retention facilities requires permissions issued by the environmental protection administration and building permissions issued by the county administration and by the municipalities.

In general, some small retention measures, such as building small retention reservoirs on rivers, reservoirs built in local terrain depressions and reconstruction of small ponds, can have negative impacts on ecosystems (destruction of valuable ecosystems, problems with fish migration, changes in ecosystems, changing ecosystems to less valuable ones, etc.). Moreover, there are collisions with agricultural production (loss of agricultural area; the possibility of excessive waterlogging of the soil etc.) (Mioduszewski and Okruszko 2016). An important collision relates to focus on retention versus focus on biodiversity protection. Planners of retention programs need to verify the location of investments in order to deal with this issue. Hydrologists engaged in the program look at a proposed measure's design thoroughly in terms of water flow and the retention targets to be achieved. Therefore, they search for a feasible and cost-effective solution and the best locations. However, the proposed locations are not necessarily valuable from the perspective of biodiversity conservation. For instance, from the point of view of flood management enlarging an existing pond can lead to a significant increase of retention potential, but it could mean enlarging an already wet area. From a biodiversity protection standpoint, installing a pond in an area scarce in terms of surface water would be more valuable.

Although the tensions between functions and sectors can cause difficulties for small retention, the implementation of the small retention programs in the forests faced relatively few problems of this type. This happy circumstance was due to two main reasons: firstly, most of the measures were applied in forests owned by the SF National Forest Holding. The dominance of one owner of land and one sector (forestry) diminishes the collisions. Secondly, the project was executed on a competitive basis. The SF National Forest Holding units or other interesting stakeholders proposed the measures and locations. It helped to eliminate proposals, a process that tends to elicit tensions. In this respect, the program is based on the no regret solutions strategy. Thirdly, stakeholders as environmental NGOs and experts were involved in the planning processes, for example in publishing a handbook on technical aspects of small retentions (CKPŚ 2009, 2008; Goździk et al. 2009; CKPŚ 2016) was consulted by seven NGOs. This is a good way to handle different points of view and conflicts in an attempt to achieve an all around win–win situation.

The Problem of up-Scaling

The small retention program in the Polish Forests, in its consecutive phases, resulted in realizing several thousand local, small-scale projects. However, from the management point of view, the program can be treated as a top-down initiative. The SF National Forest Holding is a large organization with 450 territorial units and 25,000 employees. It has a hierarchical structure with strict rules and procedures. Therefore, any new approach has to be approved by the Headquarter Directors. The program was developed at the ministerial level and was launched by the director of the SF National Forest Holding. In 1997, the director of the SF National Forest Holding approved a guiding document: Principles of planning and implementation of small retention in the State Forests, defining the concept of "small retention". In 2002, a new agreement was signed "on cooperation to increase the development of small water retention and the dissemination and implementation of pro-ecological methods of water retention", which was signed additionally by the directors of the National Fund for Environmental Protection and Water Management and the Agency for Restructuring and Modernization of Agriculture. Moreover, the small retention program relies on Regulation No. 11 of the director of the SF National Forest Holding (February 14, 1995) on improving forest management on ecological grounds and several other documents and guidelines concerning forest management. According to the current instructions, namely the Forest Management Rules (2012), the possibility of increasing retention in forests by improving functionality, restoration or construction of new drainage devices is taken into account (Liberacki et al. 2016). These documents set up the regulatory framework for realization of the small retention programs.

From this point of view, the regulatory background allowing for small retention initiatives was set by the top decision makers. At the same time, the project relied largely on bottom-up initiatives. The SF National Forest Holding territorial units needed to prepare projects and apply within a grant-like procedure. The chance to obtain financing was an obvious incentive, encouraging the territorial units to propose applications. The program was a success, with many applications. The quasi voluntary recruitment appeared effective in initiating the program and achieving the critical mass. However, this format also resulted in certain spontaneous characteristics in terms of program development. For instance, at the beginning, many projects were focused on dams on lakes simply due to the feasibility of such projects, which allowed for a quick increase of retention potential. When these options were used up, other measures became more popular. In order to supervise program consistency and the program life cycle, in the third phase the additional monitoring component was established.

Conclusions

After 1997, several small-scale retention programs in forests were introduced in Poland, reversing the earlier practice of drainage. Flood prevention is a part of a wider portfolio of the programs' benefits (forestry, agriculture, biodiversity protection).

The relative feasibility of the small retention program in terms of property rights can be attributed to a specific characteristic of the SF National Forest Holding. It is a corporation but shares many features of an organization working for the public interest. Moreover, it is a hierarchical organization; thus the decision of the headquarters to launch the programs was effectively implemented. Also, the SF National Forest Holding owns large forest areas, a crucial factor for an undertaking as demanding for space as water retention development. Although in several cases cooperation with local governments was successful, in most cases the Holding relied on its own properties. Next, the Holding was able to initiate the programs and to carry them out because of its good economic standing. Although the programs were subsidized, the initial investment was required and the Holding was able to sustain it. Finally, the hierarchical structure of the Holding was instrumental to up-scaling. At the beginning, the programs were experimental and followed a learning-by-doing method. It was later expanded and corrected. It was possible, largely, due to stability of the policy of the SF National Forest Holding.

The life cycle of the project (three phases over the course of 20 years) offers several lessons: (a) establishing an encouraging legal and institutional environment is crucial for initiating a program such as the small retention programs; (b) providing financial subsidies is a prerequisite for success; (c) a procedure with calls of proposals and application from the local branches of the Holding resulted in a search for "low hanging fruits", which helped to reach the critical mass and gather experience.

Acknowledgements Open access of this chapter is funded by COST Action No. CA16209 Natural flood retention on private land, LAND4FLOOD (www.land4flood.eu), supported by COST (European Cooperation in Science and Technology).

References

Ahlers AA, Heske EJ, Schooley RL (2016) Prey distribution, potential landscape supplementation, and urbanization affect occupancy dynamics of American mink in streams. Landsc Ecol 31(7):1601–1613. https://doi.org/10.1007/s10980-016-0350-5

Baldwin RF (2005) Vernal pools: critical habitat. Front Ecol Environ 9:471–477

Baldwin RF, Calhoun AJK, de Maynadier PG (2006) The significance of hydroperiod and stand maturity for pool-breeding amphibians in forested landscapes. Can J Zool 84:1604–1615

Brzeziński M, Natorff M et al (2012) Numerical and behavioral responses of waterfowl to the invasive American mink: a conservation paradox. Biol Conserv 147:68–78

CKPŚ (2008) Podręcznik Wdrażania Projektu—Wytyczne do realizacji zadań i obiektów małej retencji na ternach nizinnych: Zwiększanie możliwości retencyjnych oraz przeciwdziałanie powodzi i suszy w ekosystemach leśnych na terenach nizinnych (Project management tutorial—guidelines for realising small retention project on the lowlands: increasing the retention potential and mitigating flood and drought in lowland forest ecosystems). CKPŚ, PGL Lasy Państwowe, Warszawa

CKPŚ (2009) Podręcznik Wdrażania Projektu—Wytyczne do realizacji zadań i obiektów małej retencjina ternach nizinnych: Zwiększanie możliwości retencyjnych oraz przeciwdziałanie powodzi i suszy w ekosystemach leśnych na terenach nizinnych (Project management tutorial—guidelines for realising small retention project in the mountains: increasing the retention potential and mitigating flood and drought in highland forest ecosystems). CKPŚ, PGL Lasy Państwowe, Warszawa

CKPŚ (2016) Podręcznik Wdrażania Projektu—Wytyczne do realizacji zadań i obiektów małej retencji i przeciwdziałania erozji wodnej: Kompleksowy projekt adaptacji lasów i leśnictwa do zmian klimatu—mała retencja oraz przeciwdziałanie erozji wodnej na terenach nizinnych i górskich (Project management tutorial—guidelines for realising small retention project and prevent erosion: a complex project on the adaptation of forests to climate change, increasing small retentions and prevent erosion in lowlands and highlands). CKPŚ, PGL Lasy Państwowe, Warszawa

Di Mauro D, Hunter ML Jr (2002) Reproduction of amphibians in natural and anthropogenic temporary pools in managed forests. For Sci 48:397–406

Fournier MC, Larrue M et al (2016) Flood risk mitigation in Europe: how far away are we from the aspired forms of adaptive governance? Ecol Soc 21(4):49–59. https://doi.org/10.5751/ES-08991-210449

Goździk M, Guzek K et al (2009) Podręcznik wdrażania projektu—Wytyczne do realizacji małej retencji w górach: Przeciwdziałanie skutkom odpływu wód opadowych na terenach górskich. Zwiększenie retencji i utrzymanie potoków oraz związanej z nimi infrastruktury w dobrym stanie (Project management tutorial—Guidelines for realisation of small retention program on the highland forested areas. Mitigation of the impact of precipitation flow on the highland areas. Increasing retention and management of running waters and related infrastructures). CKPŚ, PGL Lasy Państwowe, Warszawa

Hanowski J, Danz N, Lind J (2006) Response of breeding bird communities to forest harvest around seasonal ponds in northern forests, USA. For Ecol Manag 229:63–72. https://doi.org/10.1016/j.foreco.2006.03.011

JASPERS (2009) Wstępna ocena zbiorników realizowanych w ramach projektu: Przeciwdziałanie skutkom odpływu wód opadowych na terenach górskich. Zwiększenie retencji i utrzymanie potoków oraz związanej z nimi infrastruktury w dobrym stanie (Preliminary analysis of retention sites realised within the project: mitigation of the impact of precipitation flow on the highland areas. Increasing retention and management of running waters and related infrastructures). JASPERS, PGL Lasy Państwowe, Warszawa

Jelonek M, Engel J et al (2008) Wstępna ocena projektu: Przeciwdziałanie skutkom odpływu wód opadowych na terenach górskich. Zwiększenie retencji i utrzymanie potoków oraz związanej z nimi infrastruktury w dobrym stanie (Preliminary assessment of the project: mitigation of the impact of precipitation flow on the highland areas. Increasing retention and management of running waters and related infrastructures). CKPŚ, PGL Lasy Państwowe, Warszawa

Juszczak R, Kędziora A, Olejnik J (2007) Assessment of water retention capacity of small ponds in agricultural-forest catchment in Western Poland. Pol J Environ Stud 16(5):685–695

Koseoglu N, Moran D (2014) Review of current knowledge. Demystifying natural water retention measures (NWRM). Foundation of Water Research, Marlow, Bucks, pp 3–15

Kowalczak P (2002) Hierarchia potrzeb obszarowych małej retencji dla wojewodztwa zachodniopomorskiego (Hierarchy of the local demand for small retention in the Western Pomeranian Region). IMGW, Warszawa

Kowalewski Z (2008) Actions for small water retention undertaken in Poland. J Water Land Dev 12:155–167. https://doi.org/10.2478/v10025-009-0012-y

Kundzewicz ZW, Dobrowolski A et al (2012) Floods in Poland. In: Kundzewicz ZW (ed) Changes in flood risk in Europe. IAHS Press

Liberacki D, Korytowski M et al (2016) Efekty realizacji programu małej retencji w lasach na przykładzie dwóch nadleśnictw obszarów nizinnych (Effects of realisation the small retention program in forests, on the example of two lowland forestry). Rocznik Ochrona Środowiska 18:428–438. ISSN 1506-218X

Miler AT (2015) Mała Retencja Wodna w Polskich Lasach Nizinnych. Infrastruktura i Ekologia Terenów Wiejskich (Infrastruct Ecol Rural Areas) 4(1):979–992. http://dx.medra.org/10.14597/infraeco.2015.4.1.078

Mioduszewski W (2014) Small (natural) water retention in rural areas. J Water Land Dev 20(1):19–29. https://doi.org/10.2478/jwld-2014-0005

Mioduszewski W, Okruszko T (eds) (2016) Naturalna, mała retencja wodna—Metoda łagodzenia skutków suszy, ograniczania ryzyka powodziowego i ochrona różnorodności biologicznej (Natural small water retention—a method for lowering the effects of droughts and risk of floods, and increasing biodiversity). Podstawy Metodyczne, Globalne Partnerstwo dla Wody. http://gwppl.org/data/uploads/dokumenty/naturalna_mala_retencja_mioduszewski_okruszko.pdf

Mioduszewski W, Pierzgalski E (eds) (2009) Zwiększenie możliwości retencyjnych oraz przeciwdziałanie powodzi i suszy w ekosystemach leśnych na terenach nizinnych (Increasing retention capacity and flood and drought prevention in lowland forest ecosystems). CKPŚ, PGL Lasy Państwowe, Warszawa, pp 1–73

Nordström M, Högmander J et al (2002) Variable responses of waterfowl breeding populations to long-term removal of introduced American mink. Ecography 25:385–394. https://doi.org/10.1034/j.1600-0587.2002.250401.x

Pierzgalski E, Wolicka M, Niemtur S (2017) The environmental aspects of water management in mountain forests—Polish experiences. EFI technical report 101. https://link.springer.com/chapter/10.1007/978-3-319-57946-7_2

Puttock A, Graham HA et al (2017) Eurasian beaver activity increases water storage, attenuates flow and mitigates diffuse pollution from intensively-managed grasslands. Sci Total Environ 576:430–443

Schüler G (2006) Identification of flood-generating forest areas and forestry measures for water retention. For Snow Landsc Res 80(1):99–114

Tyszka J (2009) Estimation and economic valuation of the forest retention capacities. J Water Land Dev 13a:149–159

Wegner S (1999) A review of the scientific literature on riparian buffer width, extent and vegetation. Publication of Office of Public Service and Outreach, Institute of Ecology, University of Georgia, Athens, Georgia, pp 1–59

Whitaker DM, Montevecchi WA (1997) Breeding bird assemblages associated with riparian, interior forest, and nonriparian edge habitats in a balsam fir ecosystem. Can J For Res 27:1159–1167

Zając S (2004) Legal and financial instruments in Polish forest policy. In: Schmithüsen FJ, Trejbalová K, Vančura K (eds) Legal aspects of European forest sustainable development. Conference proceedings. https://doi.org/10.3929/ethz-a-005977041

Piotr Matczak studied sociology, culture science, and environmental management in Poland and the Netherlands. He is the Chair for Local and Regional Governance at the Institute of Sociology, Adam Mickiewicz University in Poznan (Poland). His research covers governance and institutional aspects of public policies concerning natural disasters, water management, climate change, and nature conservation.

Viktória Takács studied biology and environmental management in Hungary and in the Netherlands, obtained a Ph.D. in Poland. She participated in several projects concerning ecology, ecosystem services, forest biodiversity monitoring, and nature conservation.

Marek Goździk studied forestry and environmental engineering in Wroclaw (Poland). He holds a doctorate in river engineering. Currently, he coordinates small retention projects in State Forests (CKPŚ).

Chapter 4
Commentary: A (Mostly) Hydrological Commentary on the Small Retention Programs in the Polish Forests

Martyn Futter

From a hydrological perspective, FRM is simple: reduce the height of the flood peak and the flood risk is reduced. In reality, a huge number of biophysical and societal factors complicate the process. The Polish forest small retention program is unique in its manner of addressing these factors. The contribution of Matczak et al. (this volume) is especially welcome as it contributes to the English language literature on practical measures for upstream water retention so as to reduce downstream flood risk. While there are a few publications in English about the Polish programme (e.g., Juszczak et al. 2007; Kowalewski 2008), the voluminous Polish language literature appears to be a treasure trove of useful information for forest managers interested in water retention as well as water managers who are interested in the flood reduction potential of the forest landscape.

There is an apparent contradiction between forest production and flood control. As early as the 1700s, it was argued that site drainage would improve forest productivity (Glacken 1967, p. 488). Measures such as ditching to remove water from poorly drained lands are recognised as among the most cost-effective silvicultural practices for increasing forest yield (Skaggs et al. 2016). Between 1945 and the collapse of the Soviet Union, close to 3% of the Polish land area was drained, primarily to increase timber production. This led to lower groundwater levels and the loss or degradation of many forest wetlands (Matczak et al. this volume). As the purpose of drainage is to move water off the land, it is likely to increase the height of the flood peak and to exacerbate flooding (e.g., Robinson et al. 2003). So long as forest lands were managed primarily for production, this tradeoff was apparently met with tacit acceptance. In the 1990s, with a reduced focus solely on timber production values, the tradeoffs associated with forest drainage began to be scrutinized more closely (Mårald et al. 2017). For example, new drainage of forest lands in Finland declined substantially in the early 1990s over concerns about biodiversity (Peltomaa 2007).

M. Futter (✉)
Department of Aquatic Sciences and Assessment, Section for Geochemistry and Hydrology, Swedish University of Agricultural Sciences, Uppsala, Sweden
e-mail: martyn.futter@slu.se

T. Hartmann et al. (eds.), *Nature-Based Flood Risk Management on Private Land*,
https://doi.org/10.1007/978-3-030-23842-1_4

Matczak et al. (this volume) note that drainage of forest lands in Poland decreased at the same time as national priorities changed with the collapse of the Soviet Union.

Once forest management no longer focused exclusively on timber production, a multifunctional perspective could be adopted and other management goals increased in importance (Farrell et al. 2000). Often, these new goals were related to biodiversity or recreational values, but the importance of water management was also recognised in some locations. With their typically high infiltration capacities and potential for water purification, forests are recognised almost universally as high quality fresh water sources (Neary et al. 2009). By returning moisture to the atmosphere through evapotranspiration, forests are also an important part of the global hydrological cycle (Launiainen et al. 2014). However, the overall link between floods and forests is a matter of ongoing debate (CIFOR 2005). Despite the broader debate about the role of forests in the hydrological cycle, it seems clear from basic hydrological principles that slowing down runoff will flatten out the hydrograph and reduce the height of the flood peak.

The study of Matczak et al. (this volume) raises a number of questions. First of all, to what extent can small retention features enhance the flood risk reduction potential of existing forests? Second, what are the key institutional factors that led to the success of the Polish programme, and are they present throughout Europe? The third question has two parts—what is the optimal arrangement of water retention features in the forest landscape, and how well can human and non-human ecological engineers implement these arrangements?

Using a combination of landscape features and engineering approaches to retain water, as is done in the Polish forest small retention programme, will make a number of contributions to flood reduction. Up to the point at which all available water storage is filled, the downstream flow of water will be slowed and the hydrograph will be flatter than would be expected in the absence of water storage features. It is also likely that water retention features will contribute to increased groundwater recharge, offsetting some of the negative effects of earlier forest drainage. Furthermore, the interception and transpiration of the forest canopy can significantly reduce the fraction of precipitation that eventually contributes to runoff (e.g., Calder 1990).

As noted by Matczak et al. (this volume), small water retention measures are not a panacea for flood risk reduction. The scientific literature is in overwhelming agreement that a certain scale of floods will exceed landscape retention capacity, regardless of the forest management measures employed (e.g., Soulsby et al. 2017). However, increasing evidence suggests that targeted forest management measures can reduce the frequency and/or severity of small floods.

It is likely that one key factor of the success of the Polish small water retention programme is the forest ownership structure. Forests are often owned or managed by the state, which facilitates a landscape-scale approach to water management. Private forest owners are very constrained in their decision-making, and most decisions are made or approved by a state actor (Nichiforel et al. 2018). As a single actor has the ability to make decisions about land management across a large spatial area, ponds and other water retention features can be constructed where they are most appropriate from a hydrological perspective. This might not be possible in countries where forest ownership is dominated by a large number of smallholdings, or where forest owners are unwilling to accept loss of future timber revenue associated with the construction of ponds and other water retention features.

The optimal arrangement of water retention features in the landscape is not always apparent. In Poland, the same institutional constraints that would likely exist in countries with forest smallholdings do not apply in the same way. However, the necessary tools for optimal location of retention features may not be fully utilized. Salazar et al. (2012) report a modelling study of pond effectiveness for water retention in a series of catchments across Europe. Matczak et al. (this volume) note that similar studies have been conducted in Poland, but it seems that other modelling tools could give more useful results than the ones they present. Specifically, GIS models based on high resolution LiDAR data have a great deal of potential for identifying candidate water retention sites (Collentine and Futter 2018). Much, if not all, of Poland is now covered by a suitable LiDAR data set, and this could be exploited more effectively.

One open question is the role of beavers in the Polish forest landscape. Matczak et al. (this volume) allude briefly to this issue, but it may be more prominent in the future (e.g., Gorczyca et al. 2018). Beavers are ecological engineers that build dams to create flooded habitat. Beaver ponds can function in a different ecological and hydrochemical manner than small artificial ponds (Ecke et al. 2017). These differences should be explored further as beavers continue to recolonize the Polish forest.

We can learn a number of lessons from this case study in particular and the Polish forest small water retention program in general. Perhaps the most important of these is that multifunctional forestry can work. Forest harvest, expressed as roundwood removal rates, has continued to increase (Eurostat, "roundwood removals by type and assortment") at the same time as pond construction and forest water retention capacity have increased. Equally important to the success of the Polish programme are the institutional and governance structures that facilitate the adoption and construction of water retention features. This may not be possible in jurisdictions dominated by forest smallholdings or where the capacity for top-down decision-making is limited. Two further lessons may not be as important but equally relevant to the question of flood risk reduction measures on private land. The first has to do with the incomplete uptake of the available technology. Models are available for quantifying landscape-scale water retention capacity, and new data, specifically high-resolution LiDAR surveys can support these initiatives. The second is how to better integrate human and non-human ecological engineers in production landscapes. Beavers are perceived as damaging to the forest economy but can be an important contributor to forest multifunctionality as their dams and ponds increase water holding capacity and can act as biodiversity hotspots.

The overarching message for my discipline and my colleagues is to learn from the Polish experience. The small water retention programme works, and, with appropriate adaptation, it could be implemented elsewhere in Europe.

Acknowledgements Open access of this chapter is funded by COST Action No. CA16209 Natural flood retention on private land, LAND4FLOOD (www.land4flood.eu), supported by COST (European Cooperation in Science and Technology).

References

Calder IR (1990) Evaporation in the uplands. Wiley

CIFOR (2005) Forests and floods: drowning in fiction or thriving on facts? Forest Perspectives 2, RAR Publication 2005-3

Collentine D, Futter MN (2018) Realising the potential of natural water retention measures in catchment flood management: trade-offs and matching interests. J Flood Risk Manag 11(1):76–84

Ecke F, Levanoni O, Audet J, Carlson P, Eklöf K, Hartman G, McKie B, Ledesma J, Segersten J, Truchy A, Futter M (2017) Meta-analysis of environmental effects of beaver in relation to artificial dams. Environ Res Lett 12(11):113002

Farrell EP, Führer E, Ryan D, Andersson F, Hüttl R, Piussi P (2000) European forest ecosystems: building the future on the legacy of the past. For Ecol Manag 132(1):5–20

Glacken CJ (1967) Traces on the Rhodian Shore: nature and culture in Western thought from ancient times to the end of the eighteenth century, vol 170. University of California Press

Gorczyca E, Krzemień K, Sobucki M, Jarzyna K (2018) Can beaver impact promote river renaturalization? The example of the Raba River, southern Poland. Sci Total Environ 615:1048–1060

Juszczak R, Kędziora A, Olejnik J (2007) Assessment of water retention capacity of small ponds in Wyskoć agricultural-forest catchment in western Poland. Pol J Environ Stud 16(5)

Kowalewski Z (2008) Actions for small water retention undertaken in Poland. J Water Land Dev 12:155–167

Launiainen S, Futter MN, Ellison D, Clarke N, Finér L, Högbom L, Laurén A, Ring E (2014) Is the water footprint an appropriate tool for forestry and forest products: the fennoscandian case. Ambio 43(2):244–256

Mårald E, Sandstrom C, Nordin A (2017) Forest governance and management across time: developing a new forest social contract. Routledge, Abington

Neary DG, Ice GG, Jackson CR (2009) Linkages between forest soils and water quality and quantity. For Ecol Manage 258(10):2269–2281

Nichiforel L, Keary K, Deuffic P, Weiss G, Thorsen BJ, Winkel G, Avdibegović M, Dobšinská Z, Feliciano D, Gatto P, Mifsud EG (2018) How private are Europe's private forests? A comparative property rights analysis. Land Use Policy 76:535–552

Peltomaa R (2007) Drainage of forests in Finland. Irrig Drain: J Int Comm Irrig Drain 56(1):151–159

Robinson M, Cognard-Plancq AL, Cosandey C, David J, Durand P, Führer HW, Hall R, Hendriques MO, Marc V, McCarthy R, McDonnell M (2003) Studies of the impact of forests on peak flows and baseflows: a European perspective. For Ecol Manage 186(1–3):85–97

Salazar S, Francés F, Komma J, Blume T, Francke T, Bronstert A, Blöschl G (2012) A comparative analysis of the effectiveness of flood management measures based on the concept of" retaining water in the landscape" in different European hydro-climatic regions. Nat Hazards Earth Syst Sci 12(11):3287–3306

Skaggs RW, Tian S, Chescheir GM, Amatya DM, Youssef MA (2016) Forest drainage. In: Amatya DM, Williams TM, Bren L, de Jong C (eds) Forest hydrology: processes, management and assessment. CABI, pp 124–140

Soulsby C, Dick J, Scheliga B, Tetzlaff D (2017) Taming the flood—how far can we go with trees? Hydrol Process 31(17):3122–3126

Martyn Futter is a hydrologist with a strong interest in quantitative hydrology and the social dimensions of water. He has worked extensively with forest water issues in Sweden and has developed a rainfall runoff model that has found use in North America, Europe, Asia and Africa.

Chapter 5
Commentary: Small Retention in Polish Forests from a Forest Management Perspective—Copying of Existing Could Be Right Path

Marijana Kapović Solomun

The case has shown how a distinct approach in natural disaster mitigation and resilience using small retention facilities in public forests can be effective. These insights also offer a new perspective on how public forests and forest lands can be managed to mitigate extreme events like floods, droughts, or wildfires, and contribute to resilience on climate change at the same time. Afforestation has been perceived as the most common sustainable measure used for forest and land management in the most countries, specifically developing one (Kapović Solomun et al. 2018). However, another possible approach for mitigating extreme events using small water retention facilities will be argued here. There are numerous similarities and challenges in the implementation of small retention facilities that can be recognized by neighboring, particularly post-communist countries, but there also possible solutions of how to cope with them. Natural resource dependent countries need to start to form consistent implementation of legislative framework and policies, where government leadership is crucial. Knowing that the main abiotic disturbances in Europe are fire, wind, flooding and drought (Flannigan et al. 2000; Moriondo et al. 2006) makes information relating to climate change and the human-induced factors even more valuable. Disaster prevention and mitigation has been a priority for the last several decades, but the question is this: How do governments find the trade-offs and the right path for the future and provide sustainability for forests, biodiversity, land and human demands at the same time? This case is specific to Poland and provides the possibility to introduce new natural and technical measures in forestry to prevent and mitigate flood, drought and wildfire. At the same time, this is an example relevant to countries with a prevalence of public forests and centralized forest management: What can be done to mitigate natural disasters trough the forestry sector? Small retention facilities are a step forward in improving the prevention and mitigation of such extreme events, with, unfortunately, limited evidence of effectiveness so far. Using forests for successful and large-scale prevention and mitigation requires con-

M. Kapović Solomun (✉)
Faculty of Forestry, University of Banja Luka, Banja Luka, Bosnia and Herzegovina
e-mail: marijana.kapovic-solomun@sf.unibl.org

T. Hartmann et al. (eds.), *Nature-Based Flood Risk Management on Private Land*,
https://doi.org/10.1007/978-3-030-23842-1_5

sistent policy-directed solutions and strong cooperation among governments, forest scientific agencies, local communities, private landowners, private forest owners and NGOs.

Forest Ownership and Forest Stewardship

Forest ownership is a sensitive subject, particularly in regards to implementation of new measures like small retention facilities. When the new measures are brought up and consider public forests with centralized forest management, the various stakeholders quickly voice numerous challenges and conflicts. Knowing that, what can be realistically expected in regards to private forests and privately owned land, where ownership and land rights are far more delicate? Private forest owners are an important target group in terms of prevention and mitigation of extreme events, notably in the countries with a larger percentage of private forests. How is it possible to involve land and forest owners and convince them to participate in the utilization of small retention facilities and harmonize all interests within their community for everyone's benefit? These are important questions for the future.

Introduction of new techniques, measures, and approaches (like small retention facilities), will influence private property and need efficient public-private partnership. Private landowners usually look into costs, benefits and possible effects any approach may have, and it all comes down to one question: Why would I want to pay for that? Normally, private sector engagement in adaptation assures greater investments and services in core development sectors (Biagini and Miller 2013). Community involvement and private-public partnerships can be powerful mechanisms in coping with floods (Chartres and Noble 2015) but also other natural disasters. Transformational change and policy vision are a prerequisite for a successful implementation of science-driven natural or technical measures. Obviously, extreme events will become more severe and frequent. In this case, small retention facilities could be "a bright spot" for forest managers and policy makers in other countries who will use this example and incorporate it in existing management of forest resources. Land and forest management carries a great deal of importance in addressing current risks that are accentuated by climate change impacts; this is also stated in IPCC reports (2012). However, sustainable forest management and the prevention and mitigation of floods, droughts, and wildfires, are still not perceived as an important endeavor in developing and post-conflict countries with weak socio-economic conditions (Kapović Solomun et al. 2018). Under these circumstances, the forest- and land-related legislative framework is usually adequate and includes "sustainability", but the resources for implementation are limited. Likewise, the pros and cons of centralized forest management are an argument worth consideration within country-specific circumstances. For instance, centralized forest governance and prevalence of public forests in countries with a high level of corruption and weak implementation of legislative frameworks does not usually lead to good forest stewardship. Sometimes, such centralized power can serve as an open channel for corruption among forest managers; this often leads to the exaggerated use and future decrease of forest quality, decreased interception and evapotranspiration, soil storage capacity, all of which increases the possibility for flooding and landslides as well. Furthermore, fire is understood as a natural factor, but wild fires are often threats to forests, public safety and property (Martell 2007), endangering biodiversity and land. This phenomenon is also frequently caused by humans and made more likely due to climate change. With all of this in mind, small retention facilities in the forests are potential solutions to these problems that need to be supported with continuous education and by raising awareness in communities to prevent the unnatural causes of wildfires.

Role of Local Stakeholders

Local governments often play a vital role in modulating the outcomes of sustainable forest and land initiatives (Kemp et al. 2005). Several example cases have shown that decreasing the influence and role of local community institutions has resulted in lower success rates in forest management (Ostrom 2009; Dressler et al. 2010). A greater level of forest degradation is usually evident in the vicinity of rural households (Kapović et al. 2013) where people perceive public forests and centralized forest governance less as communal property, particularly in developing regions with weak economies. Generally, one of the shortcomings of many land- and forest-related studies is the exclusion of the views and experiences of the landowners and land users, particularly in forest-dependent communities (Andersson et al. 2011). Natural flood management and ecosystem based approaches can manage flood risks, impacts and vulnerability (Gill et al. 2007; Munang et al. 2013), but it should be developed in close cooperation with local communities. A connected (involved) community (local, national, regional and international) is likely to make major headway in understanding the role of small retention facilities in recent floods, drought and wildfires. Involving stakeholders who are dependent on land, forests, or agriculture proves influential in their understanding of natural disasters, opportunities for prevention and mitigation, and their implication and position within those processes. Encouragement and active inclusion of private land and forest owners will increase the chances for success of any natural or technical measure, principally on private land, but also to prevent some natural disasters like wildfires. This case is also an interesting example of wide participation of different groups of stakeholders in a small retention facilities program related to public forests. However, it would be intriguing to see a community's response if small retention facilities are planned on private land or in private forests. Implementation of coherent national forest policy focused on sustainability and multi-functionality is a good test for even centralized forest management and public forests.

Why Small Retention Facilities?

A small retention program in the Polish forests comprises natural and technical measures for lowland and upland forests. The primary advantages are mainly related to the natural retention facilities, while adverse effects are related to the technical small retention facilities, especially in terms of biodiversity preservation. A recent study (Huang et al. 2018) suggests that rainfall and the forest landscape are pivotal factors triggering flood event alterations in lower return periods, which flood event dynamics in higher return periods are attributed to hydrological regulations of water infrastructures. Water balance, and thus retention of forested landscapes, is, to a large extent, controlled by forest ecosystems (Döll 2009), which include surface runoff as determined by evaporation, transpiration and water flow routing (Eisenbies et al. 2007). Small water retention facilities under forests contribute to increased evaporation and transpiration and lead to fewer flood events with a "keep the rain where it falls" approach. Hence, if evaporation and transpiration are predicted to decrease, that can increase surface runoff (Schlesinger and Jasechko 2014) and, consequently, floods. Still, small reservoirs may lose up to 50% of their stored volume due to evaporation in many regions and the high ratio of surface/volume area. Evaporation constitutes a major component of the water balance in the reservoirs and may significantly decrease flood events (Ashraf et al. 2007). In forests, the interception of precipitation and its loss by evaporation is typically 10–35% of precipitation (Wang et al. 2007). This interception loss in the forests is compatible with transpiration and evaporation less than 65–90% (Schlesinger and Jasechko 2014). Typically, forests and forest land increases evapotranspiration and tends to decrease the number of flood events (Ryu et al. 2011) and droughts.

Opportunities

Numerous public forest companies within the Europe have earned a Forest Stewardship Certificate (FSC) that confirms that the forest is managed with the intent of preserving the natural ecosystem and benefitting the lives of local people and workers all while ensuring it sustains economic viability. Forest management certification helps protect the people, plant and animal species that live in and around, and depend upon, the forest. Following a brief initial pre-assessment, the evaluation process consists of an in-depth review of forest management processes and their environmental, social, and economic impact, through defined criteria and indicators. FSC forest management certification is valid for five years and subject to annual checks whether FSC requirements are continuously met. It is a very competitive certificate and requires accomplishment in many areas of forest management, particularly whether a particular management maintains forests' biodiversity, productivity, regeneration capacity, vitality and their potential to fulfill, now and in the future, relevant ecological, economic and social functions at all levels. Small retention facilities as a positive practice should be introduced to the existing FSC criteria and indicators needed to sustain public forest management, as their implementation serve both a social and ecological function and mitigate natural disasters at the same time. Stakeholders' engagement in a participative approach can present another opportunity back to back with close cooperation between science study and practice in the application of new approaches and techniques.

Final Remarks

Small retention facilities are an exceptional example of how public forests can be managed to provide sustainability and multi-functionality and to mitigate floods, drought and wildfire at the same time. It could become an ideal model for countries with prevalence of public forests and centralized forest management. Effective stakeholder engagement, applying participatory approaches to create and implement small retention facilities is shown as successful example of adaptive public forest management. The biggest challenge to this approach is finding incentives and trade-offs for private land and forest owners so that small retention facilities are also used on private land and private forests. Furthermore, even though small retention facilities in Polish forests are a great example of how natural disasters can be mitigated through the forest management, it is challenging to make a predictive evaluation, since outcomes will very much depend on country-specific circumstances. Furthermore, evidence of its effects is as of yet limited. The case can be understood as a direction what else can be done for mitigation of extreme events in public forests under centralized management. Extreme events affect poverty and ecosystem vulnerability through their multiplier effect. Small retention facilities could fundamentally improve people's lives and security, in addition to mitigating floods, drought and wildfires. Impacts of new forest practices clearly showed the necessity to gain more quantitative insights prior to further evaluation of the effects.

For the future, a stronger understanding of natural disasters, impacts and relationships with land use is essential to inform decision-makers regarding the importance of forests and sustainable forest management, in prevention and mitigation of floods, drought and wildfires. Small retention facilities are also a good idea for countries in a weak socio-economic situation that suffer from floods, drought and wildfires. They can benefit from the existing knowledge, available resources and results taken from other countries that have already developed and applied new approaches and techniques in forestry, agriculture, adaptation, and mitigation. Sometimes, "copying" policy and approaches from other countries and adopting existing measures, techniques, and approaches is necessary to sustain a globally feasible path.

Acknowledgements Open access of this chapter is funded by COST Action No. CA16209 Natural flood retention on private land, LAND4FLOOD (www.land4flood.eu), supported by COST (European Cooperation in Science and Technology).

References

Andersson E, Brogaard S, Olsson L (2011) The political ecology of land degradation. Annu Rev Environ Resour 36:295–319. http://www.annualreviews.org/doi/10.1146/annurev-environ-033110-092827

Ashraf M, Kahlown MA, Ashfaq A (2007) Impact of small dams on agriculture and groundwater development: a case study from Pakistan. Agric Water Manag 92:90–98

Biagini B, Miller A (2013) Engaging the private sector in adaptation to climate change in developing 6 countries: importance, status, and challenges. Climate Dev 5:242–252. https://doi.org/10.1080/17565529.2013.821053

Chartres CJ, Noble A (2015) Sustainable intensification: overcoming land and water constraints on food production. Food Secur 7:235–245. https://doi.org/10.1007/s12571-015-0425-1

Döll P (2009) Vulnerability to the impact of climate change on renewable groundwater resources: a global-scale assessment. Environ Res Lett 4:035006

Dressler W, Büscher B, Schoon M, Brockington D, Hayes T, Kull CA, Mccarthy J, Shrestha K (2010) From hope to crisis and back again? A critical history of the global CBNRM narrative. Environ Conserv 37:5–15. https://doi.org/10.1017/S0376892910000044

Eisenbies MH, Aust WM, Burger JA, Adams MB (2007) Forest operations, extreme flooding events, and considerations for hydrologic modelling in the Appalachians—a review. For Ecol Manage 242:77–98. https://doi.org/10.1016/j.foreco.2007.01.051

Flannigan MD, Stocks BJ, Wotton BM (2000) Climate change and forest fires. Sci Total Environ 262:221–229

Gill S, Handley J, Ennos A, Pauleit S (2007) Adapting cities for climate change: the role of the green infrastructure. Built Environ 33:115–133. https://doi.org/10.2148/benv.33.1.115

Huang XD, Wang L, Han PP, Wang WC (2018) Spatial and temporal patterns in nonstationary flood frequency across a forest watershed: linkage with rainfall and land use types. Forests 9(6):339

IPCC (2012) Managing the risks of extreme events and disasters to advance climate change adaptation: special report of the Intergovernmental Panel on Climate Change. Cambridge University Press, Cambridge, UK and New York, USA, p 582

Kapović Solomun M, Barger N, Keesstra S, Cerda A, Marković M (2018) Assessing land condition as a first step to achieving Land Degradation Neutrality: a case study of the Republic of Srpska. Environ Sci Policy 90:19–27. https://doi.org/10.1016/j.envsci.2018.09.014 (in press)

Kapović M, Tošić R, Knežević M, Lovrić N (2013) Assessment of soil properties under degraded forests: Javor mountain in Republic of Srpska—a case study. Arch Biol Sci 65(2):631–638

Kemp R, Saeed P, Gibson BR (2005) Governance for sustainable development: moving from theory to practice. Int J Sustain Dev. https://doi.org/10.1504/IJSD.2005.007372

Martell DL (2007) Forest fire management, current practices and new challenges for operational researchers. In: Weintraub A, Romero C, Bjorndal T, Epstein R (eds) Handbook of operations research in natural resources. Int Ser Oper Res & Manag Sci 99(3):419–509. http://dx.doi.org/10.1139/X07-210

Moriondo M, Good P, Durao R, Bindi M, Giannakopoulos C, Corte-Real J (2006) Potential impact of climate change on fire risk in the Mediterranean area. Climate Res 31:85–95

Munang R, Thiaw I, Alverson K, Han Z (2013) The role of ecosystem services in climate change adaptation and disaster risk reduction. Curr Opin Environ Sustain 5:47–52. https://doi.org/10.1016/J.COSUST.2013.02.002

Ostrom E (2009) A general framework for analyzing sustainability of social-ecological systems. Science 325(5939):419–422. https://doi.org/10.1126/science.1172133

Ryu JH, Lee JH, Jeong S, Park SK, Han K (2011) The impacts of climate change on local hydrology and low flow frequency in the Geum River Basin, Korea. Hydrol Process 25:3437–3447

Schlesinger W, Jasechko S (2014) Transpiration in the global water cycle. Agric For Meteorol 189–190:115–117

Wang D, Wang G, Anagnostou EN (2007) Evaluation of canopy interception schemes in land surface models. J Hydrol 347(3):308–318. https://doi.org/10.1016/j.jhydrol.2007.09.041

Marijana Kapović Solomun is associate professor at Faculty of Forestry at the University of Banja Luka. Her background is forestry, forest soils, land degradation and conservation. She is a member of UNCCD SPI, and founder of the Women's Association for Nature Protection, having participated in more than 20 international and national projects in environment and land related areas. She has published more than 30 papers in international or national journals.

Part III
Small Scale NWRM in the Czech Republic

Chapter 6
This Is My Land! Privately Funded Natural Water Retention Measures in the Czech Republic

Lenka Slavíková and Pavel Raška

Abstract Do landowners realize (and privately fund) natural water retention measures (NWRM) on their own land? Why? And how are they capable of assessing the hydrological and ecological effects of these measures? The Czech case study presents the story of an individual farmer who decided to invest his private resources in water retention and biodiversity enhancement while continuing his farming practices. We describe the historical and geographical background of the case study as well as the farmer's motivations and beliefs. We also discuss the scaling-up potential of the presented case. Information for the case study was gathered with a mixed-methodology approach that combined official statistical sources of cadastral records, content analyses of local media articles, a field survey and communication with the farmer during an excursion to his farm. The intention is to show the alternative when considering NWRM implementation. This alternative is heading the same direction as public subsidy schemes. Private initiatives may support (instead of undermine) public policy goals, and they are often faster and cheaper.

Introduction

Natural water retention measures (NRWM) are often initiated by river basin authorities or nature protection agencies, implemented in cooperation with multiple public bodies and supported with a public engagement processes and public funding (Rouillard et al. 2015; Hesslerová et al. 2016). Owners of the involved land represent a key stakeholder group to take into consideration. They might oppose the realization of these measures (regardless of public interest) if not properly treated.

L. Slavíková (✉)
Faculty of Social and Economic Studies, Institute for Economic and Environmental Policy (IEEP), J. E. Purkyně University, Ústí nad Labem, Czech Republic
e-mail: lenka.slavikova@ujep.cz

P. Raška
Faculty of Science, Department of Geography, J. E. Purkyně University, Ústí nad Labem, Czech Republic
e-mail: pavel.raska@ujep.cz

T. Hartmann et al. (eds.), *Nature-Based Flood Risk Management on Private Land*,
https://doi.org/10.1007/978-3-030-23842-1_6

Hereby, planners' and decision-makers' intentions are often blocked by property right arrangements (Steinhäußer et al. 2015). Such property right arrangements may vary in different legal, tenure and planning systems, but generally they pose a contradiction between the protection of owner's interests on the one hand and the right to share and protect common environmental values on the other (Hanna et al. 1996).

For this reason, when landowners themselves take the initiative and realize (privately funded) NWRM on their own land, they should gain increasing attention. Many simple situations can arise where stakeholders may show initiative. But why they do so? Do the owners require compensations? And how are they capable of assessing hydrological and ecological effects of these measures? Numerous questions arise in the scientific community when dealing with (rather random) un-coordinated private efforts in this matter.

The Czech case study presents the story of an individual farmer who decided to "restore" his glebe. We present the extent of his activities in terms of NWRM implementation and with respect to the level of coordination with public authorities as well as his proclaimed motivations and beliefs. At the end, we discuss the scaling-up potential of the presented case. In addition to these aims, the study provides an exceptional case for the post-socialist planning context, where the engagement of private actors (and their resources) in the ecosystem service provision has been a latecomer on the conservation agenda (Pavlínek and Pickles 2000; Tickle 2000).

A qualitative exploratory case study methodology was applied to obtain a complete understanding of the case (Yin 2003; Kaae et al. 2010). The research method was considered exploratory since the aim of the empirical research was to primarily investigate motives and beliefs of an individual actor pursuing public policy goals at his own expense. The goal was not to link his behaviour to an existing theory. Data for the case study was gathered from February until November 2017 through a combination of the following methods: (a) the review of publicly available records and documents (official statistical sources of cadastral records, local media articles and interviews with the farmer), (b) a field visit of the farmland, (c) un-structured personal communication with the farmer during the guided excursion on his farm. Gathered data has been manually processed and interpreted with the use of conventional content analysis to reveal existing patterns of behaviour (Hsieh and Shannon 2005).

Geographical and Historical Background of the Case Study

The area in focus is located in the northern part of the Czech Republic. The study area (Fig. 6.1) itself reaches into the western margin of the České Středohoří Mountains, a neovolcanic landscape dominated by the highest peak Milešovka (837 m a. s. l.; Raška and Cajz 2016). The unique scenery surrounding Milešovka has been subject to nature protection; in 1951, the first nature reserve of about 60 ha was established. In 1976, the Protected Landscape Area (PLA) České Středohoří of more than 106.000 ha was founded and is now protected under Czech legislation. In the broader area, there are also numerous Natura 2000 Sites defined according to the EU Habitats Directive (České Středohoří 2017).

The study area consists of meadows, forests and small water streams. Climatically, it is located in the rain shadow of the near Krušné Mountains north-west from the study area. The annual precipitation has been one third lower than the Czech average (CHMI 2017). Depending on topography, exposure climate (southern and northern orientations) and edaphic conditions, the variety of biotopes evolved under the long-term human cultivation. While the northern steep slopes are covered with mixed forests, they continue downslope to mesophilic meadows. The southern steep slopes, in turn, receive much more solar energy and are typical for rock-mantled slopes covered by forests as well as semixerophilic and thermophilic vegetation. The southern foothills and slightly undulating surfaces were used as orchards in the past for their suitable climatic conditions (see Fig. 6.1).

The territory has been settled and cultivated since medieval times. At the beginning of the 20th century, the land was dominantly used for pasturing and small-scale agriculture combined with fruit-growing. After 1948, during the era of Communism, small parcels were expropriated, meliorated and consolidated into the large blocks of arable land for intensive agricultural purposes. The reason was the general effort of central-planned agriculture to result with large block of fields in both lowland agricultural regions as well as upland historical rural regions (Bičík et al. 2001; Orsillo 2008). After the fall of Communism in 1989 and under the circumstances of transformation towards a free-market economy (cf., Hruška et al. 2015; Kupková

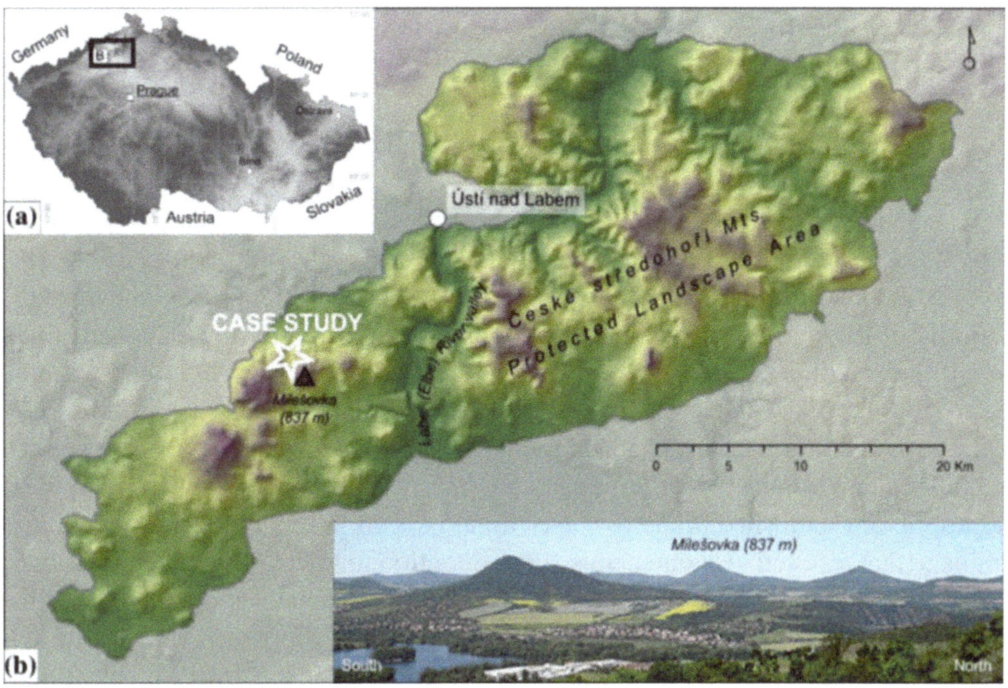

Fig. 6.1 Study area–located around the Milešovka, the highest peak of the České středohoří Mountains in NW Czechia

and Bičík 2016), the existing agriculture cooperative went bankrupt, fields were abandoned due to low yields and the spontaneous succession started. At the end of the 20th century, this succession was stopped due to renaissant management practices of a new private owner who bought the abandoned land.

Mr. Pitek's Land

Mr. Daniel Pitek was born in 1966, and he is a forest engineer by education. After the fall of Communism in 1989, he established a private business based on drywall supply using hid experiences gained in Germany. He has also ran a wood-processing company. At the turn of the millennium, Mr. Pitek decided to leave the capital Prague and to settle in the České Středohoří Mountains in the municipality of Černčice, the area where he originally came from. According to his own words, his main motive was to come back to the countryside (Respekt 2012).

First, he bought an old tavern there and established a deer preserve nearby. He observed the problematic state of the surrounding landscape ruined by the intensive agriculture and experienced first disputes with neighbors and officers because of the deer. As a result, he decided to change the status quo; but the only way forward to him was to become a landowner and to undertake changes on his own property (Respekt 2012). He started to buy the land, one parcel after another—currently he owns about 600 ha of grasslands and 30 ha of forests in dispersed patches of different size mainly surrounding the municipality of Černčice. At the beginning, he used financial resources from his existing businesses to cross-subsidize his new activities, but his goal is to run self-sustaining farms, such as sheep breading and logging. He also receives regular agricultural subsidies on the land extended under the Single Area Payment Scheme SAPS (Idnes 2015). Nowadays, he is publicly presented as the private farmer and forester (Wikipedia 2017).

On abandoned fields, Mr. Pitek firstly cut early successional pioneer species and kept strong solitaire trees and bushes as habitats for birds and other species. This practice is not common in the case of large agricultural companies that rent the majority of the land in the surrounding: "Some neighboring farmers laughed at me, when I kept solitaire trees in meadows, because you need to drive around with the tractor. For farming such a feature is seen as an obstacle" (interview with Mr. Pirek by Neovlivni 2017). Than he changed fields into grasslands as they were better suited to the local environmental conditions of the mesophilic meadows. At the southern slopes, he re-established orchards. Over time, he also grubbed out old agricultural drainage that resulted in spontaneous water retention in lower parts of the meadows. There, with the use of the excavator, Mr. Pitek created small pools to support the retention function of the terrain (Respekt 2012). Over the past 10 years, he has established about 40 small pools (Fig. 6.2). Some of them have turned into wetlands with vegetation, others are cleared out to avoid succession (information provided during the field survey). As he put it, "Building pools is my great enjoyment. But I

Fig. 6.2 One of the pools built instead of artificial drainage system on meadows at the northern footslopes of the Milešovka Mountain

do not found them for fun, they are fundamental for the surrounding countryside" (interview with Mr. Pitek by Idnes 2016).

For each small pool to be established, he only needs to obtain consent from the PLA authority and from the municipality. The pools are designed without regulation mechanisms and with respect to legislative size limits (area of 300 m^2, depth 1.5 m) in order to avoid more complicated permitting processes related to changes in land use designation. The process of establishing a pool is well described in Mr. Pitek's answer to the question: "Could you give advice on how to set up a pool to fulfill its purpose?" (interview by Idnes 2016).

1. "First, you need to reveal where the water naturally accumulates in the countryside. It is not possible to build pools at whatever places. Sometimes, old maps may help…"—in this respect he relies on the detailed knowledge of his land, and in some cases, he also checked the old plans to verify the presence and structure of the drainage structures;

2. "Than you need to deal with the administration …"—the administration process includes the permission provided by the municipal authorities and PLA authorities (see above);
3. "Than you can simply build it. My advantage is to have clay-rich sub-soil, so we excavate the hole with some profile and keep the nature to do its job. The hole firstly fills with water, than life. I do not interfere with this process any more". The excavation is mostly done during winter, when the soil is frozen, so the pool bottom can be shaped easily. The excavation work is completed with the help of the local owners of excavators. The observed speed of spontaneous succession to reach dense litoral vegetation cover usually does not exceed one or two vegetation seasons, but depends on local conditions (habitat proximity, intensity of water accumulation, suspended sediment load, etc.).

Mr. Pitek plans to build a system of 18 further large ponds (each of 2,000–3,000 m^2). However, according to his words, the permit granting procedures are very different in case of large bodies of water with controlled out-flow in comparison to the process for small pools (personal communication). The pond project has been prepared for more than one year and the construction permission has not been issued yet. Mr. Pitek seems to be frustrated with this bureaucracy.

All the undertaken and planned activities are mostly privately funded. On his land, Mr. Pitek prefers to do what he considers to be the right thing without subordinance to public authorities and/or subsidy schemes. He consults experts (hydrologists, ecologists) when necessary, but most of the time he follows his "common sense": "I create something like sponge here that absorbs the rainwater and keeps it for different uses" (interview with Mr. Pitek by Idnes 2016).

So far, no permanent monitoring or evaluation of hydrological effect is available to prove any benefits of the described effort. Having appeared in the media multiple times, Mr. Pitek's activities have recently attracted the attention of scientists; accordingly, evidence may be forthcoming in the near future. However, in terms of the ecological effects and biodiversity increase, the effect of existing pools is easy to see right now, as indicated by higher yields on meadows (reported by Mr. Pitek), improvement of the grassland diversity (monitored regularly under the legal obligation by nature conservation authorities), as well as by the presence of amphibian and bird species formerly absent at the sites (reported also by researchers from the conservation agencies and local museums). The observed effects on his land also led Mr. Pitek to criticize the existing agricultural subsidies for sending perverse incentives to land management. Within these subsidies the amount of land and the agricultural production is favoured. Water retention and environmental sustainability results in the subsidy reduction. According to his opinion (information provided during the on-site visit), such perverse incentives include, for instance, subsidies for an intensive production of crops for biofuels or the former reduction of subsidized agricultural land determined based on the area of canopy of dispersed trees on meadows and pastures (thus limiting efforts for multifunctional land uses).

Motivations and Interactions

So far, we have not provided much information about the owner's motivations. According to his statements, his goal is to restore the landscape to reflect the situation at the beginning of 19th century when farmers did not exploit the land and respected the natural limits of the land use. He uses old maps (from the first third of 19th century) to support his decisions about land use changes (Idnes 2015). His ideal, therefore, represents the cultural countryside with sustainable farming rather than wilderness. He considers current intensive agricultural practices as harmful in the long run and he feels he has been doing the right thing (Neovlivni 2017).

Water management has become his key issue due to lack of water in the area—partly due to natural conditions (rain shadow), partly due to previous mismanagement (drainage systems). Climate changes that reallocate annual rainfalls have accelerated the problem. Mr. Pitek described the countryside he saw when returning from Prague as follows: "On hillsides, there were dying trees surrounded with totally torrid landscape that slowly turned into dessert" (interview with Mr. Pitek by Neovlivni 2017). Therefore, his intention has been to reverse this situation by changed management and agricultural practices and pool restoration. Judging by his numerous statements published in mass-media interviews, Mr. Pitek is fully aware of the interconnection of farming practices and small-water cycle, and his goal is to minimize adverse effects. The increasing problem of drought (as detected by official Czech documents—see Strategy 2017) is also fully reflected: "Everyone speaks about up-coming drought, but it is already here!" (interview with Mr. Pitek by Neovlivni 2017).

Therefore, the initial motivation of pool and future pond restoration is not to contribute to flood risk mitigation, but to accumulate more rainwater in the upstream area for groundwater storage, farming and as a positive side effect for the biodiversity enhancement. Mr. Pitek is satisfied with the possibility to restore pools on his own—cheaply and fast. He is not fond of existing subsidy schemes as they frequently require increased bureaucracy (Idnes 2015).

The private effort is in line with the current management plan of the PLA České Středohoří within which Mr. Pitek's land is situated. The plan confirms a lack of water surfaces in the area, and the goal should be to establish small pools and wetlands to increase the retention capacity of the landscape (Management Plan 2017). The Czech Nature Conservation Agency and the Czech State Forest Company have re-established several pools and lakes with the use of public resources. The greatest obstacle remains the availability of land.

Therefore, the private owner and the public agencies undertake similar measures in different localities and pursue the same goals. Their mutual interactions, however, are not following official procedures. They are rather informal, based on social relationships. Mr. Pitek also serves as a PLA ranger and founded the local environmental NGO to interfere with development and restoration plans in the area. So far, the PLA has not funded any of Mr. Pitek pools from public financial resources, and there are no such plans for the future.

As apparent from abundant publicly accessible information in the press, Mr. Pitek's activities attract public attention. In 2015, he was awarded the Czech environmental price of Josef Vavroušek (Nadace partnerství 2017). He regularly speaks with media and organizes excursions for students and other persons concerned. He was the unsuccessful regional political candidate for the Green Party in 2016, but later he left the party for disagreement with its future orientation. As a result of public exposure, natural scientists intend to establish bio-monitoring at his pools to analyze ecosystem dynamics. Mr. Pitek has already built strong ties with regional institutions and scientists, such as ornitologists and botanists, who are starting to perform both the formal and informal monitoring and research activities on his land. The example of formal research is represented with periodical inventory and assessment of grasslands in the PLA České středohoří (see above), whereas informal research is mainly based on monitoring endangered species of birds, for instance. Generally, he welcomes all kinds of scientific and dissemination activities, although he does not wait for the results to proceed further.

Scaling-Up Potential?

Numerous small scale efforts of the non-public nature field protection can be detected throughout Europe and world-wide—such as land trust movements (Ruseva et al. 2016; Bastian et al. 2017), Audubon society (Merchant 2010), etc. Different non-governmental non-profit organizations (NGOs) play an important role in this process as they gain private or public resources for buying the land and change its management (including the wetland protection or re-establishment, renaturalization of small water streams, etc.). When describing their activities, authors are often concerned with transparency (Adams and Moon 2013) and sustainability of their practices, especially when day-to-day management of gained land incurs additional costs (Pasquini et al. 2011).

Initiatives by individual private owners using their own resources for water retention or biodiversity enhancement, for example, are randomly described. Reasons might be that their activities are not considered special or/and they often do not seek for the public or academic attention. This situation is in sharp contrast with the enormous effort invested in the promotion of participatory governance schemes, where getting people "on board" is the main challenge. More cases of philanthropic (or not so profit-oriented) landowners would show us that, under some circumstances, private interest may be used to increase water retention, biodiversity, etc. Furthermore, if you own the land, whatever you do is fast and cheap (although non-participatory and to some extent non-expert).

In the Czech Republic, as a post-socialist country, environmentally concerned philanthropism came back to life after 1989, and is still in its infancy. Although a growing land trust movement is detectable (CSOP 2018) and a few other private-based efforts in biodiversity conservation exist (Ceska pozice 2012), the Pitek case is exceptional in terms of its extent and the hydro-ecological focus.[1] His aim to reconcile farming with water retention and biodiversity enhancement is in sharp contrast with agricultural practices prevailing throughout the Czech Republic, where the large share of agricultural land is rented to large companies and tends to degrade (see Sklenička 2016 for further evidence).

According to our opinion, individual cases, such as the one presented in our chapter, provide two challenges. First, they facilitate understanding of the possible individual motivations of land-owners, the mechanisms of their actions and constraints they face in environmental management. Understanding these issues seems to be crucial for targeted (tailored) and effective communication of NWRM implementation under different land ownerships. To reveal these individual motivations, institutional mechanisms and constraints, we propose an analytic framework that addresses the permissive and productive conditions to perform the NRWM measures in various contexts (see Fig. 6.3).

Second, and related to the previous, unlike the organized efforts of various NGOs that are aimed at single issues (mostly but not solely biodiversity) and that frequently don't consider economic valuations as the primary criterion for environmental management, the presented case shows that individual landowners may consider environmental measures in a more complex way (water retention as well as biodiversity) and in a framework of economic sustainability since they take financial responsibility for any effects and side-effects of their action.

It is not our intention to declare that more landowners like Mr. Pitek would solve our problems with NWRM enforcement and financing. As he put it: "Of course, everything I did was possible due to incomes from my drywall supply and wood processing business. Also my entire family supports me. Otherwise I would not be able to manage it" (quote from Respekt 2012)—this shows that private initiatives may support (not only undermine) public policy goals (compare different theoretical approaches in Slavíková et al. 2010). The rationale to promote and even establish private initiatives is mostly rooted in their potential to overcome land-use conflicts and improve conservation efforts on fragmented land through coordinated action. Another rationale lies in private landowner's ability to overtake the responsibility to sustain the NWRM or other environmental measures. To allow for upscaling, such direction will, however, make it necessary (i) to explore the various territorial and legal contexts (areas under conservation, tenure schemes etc.) for privately-funded NWRM, and (ii) to reconsider the current agricultural subsidy schemes in order to stress their complex dedication, thus debunking the misconception that they shall support the sole productive functions of agricultural land.

[1]Researchers have identified a few other efforts in the Czech Republic, but these (i) are still in their beginnings with no published data, and (ii) represent the participatory rather than the bottom-up approach initiated by land-owners. These examples, for instance, include the current (re-)establishment of pools, ponds and greenery in the extensive agricultural land near the city of Liberec (Northern Czech Republic). In this case, the regional authorities looked for land threatened by droughts and have found a land-owner with ca. 1000 ha land, who—upon negotiations—was keen to realize the plans using both subsidies and his own financial sources.

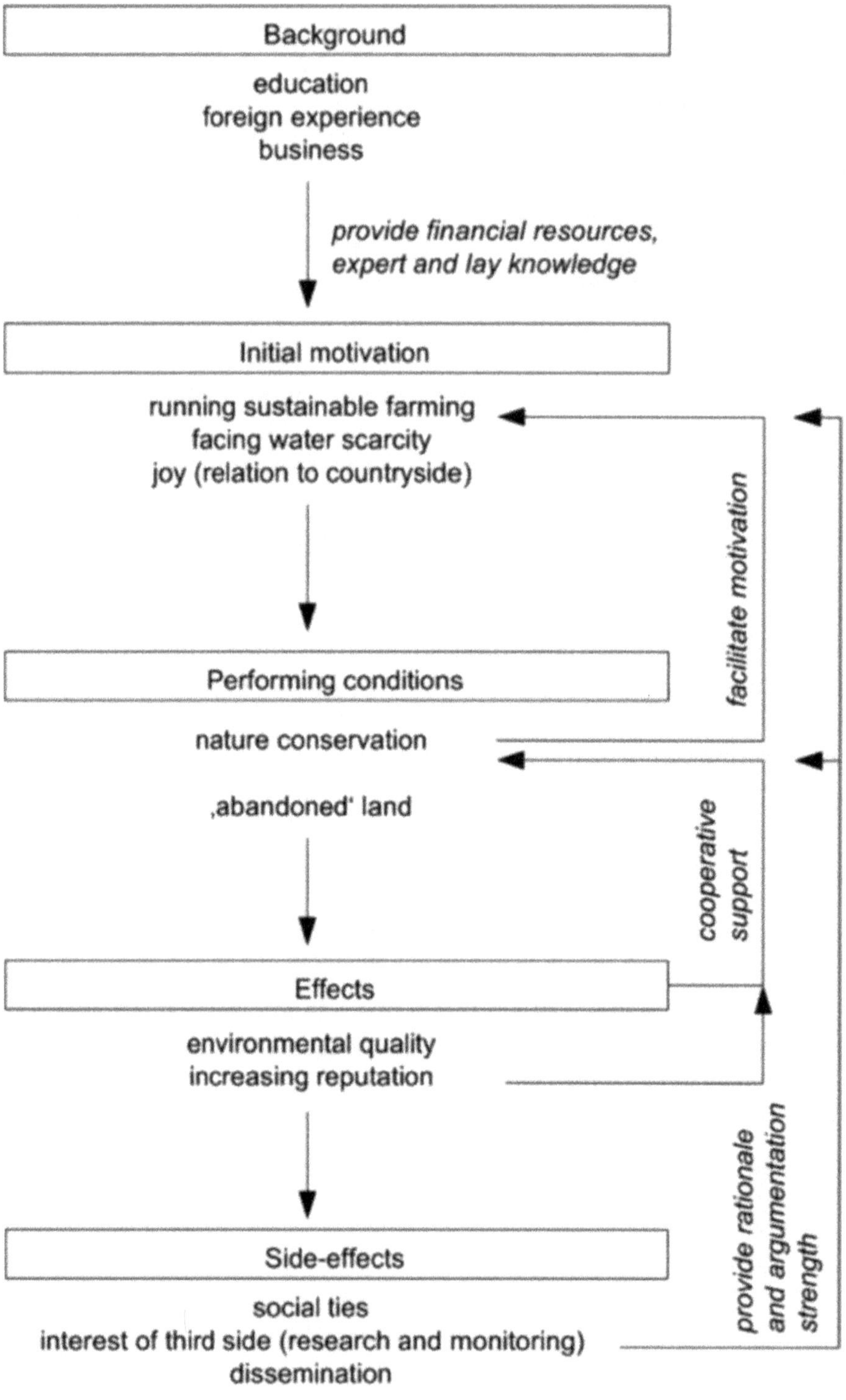

Fig. 6.3 Tentative framework to assess the feedback between motivation for and effects of the environmental action based on the case study

Acknowledgements We thank the Operational Programme Research, Development and Education of the Czech Republic for financing the project Smart City—Smart Region—Smart Community (grant number: CZ.02.1.01/0.0/0.0/17_048/0007435) that led to the present book chapter.

Open access of this chapter is funded by COST Action No. CA16209 Natural flood retention on private land, LAND4FLOOD (www.land4flood.eu), supported by COST (European Cooperation in Science and Technology).

References

Adams VM, Moon K (2013) Security and equity of conservation covenants: contradictions of private protected area policies in Australia. Land Use Policy 30(1):114–119. https://doi.org/10.1016/j.landusepol.2012.03.009

Bastian CT, Keske CM, McLeod DM, Hoag DL (2017) Landowner and land trust agent preferences for conservation easements: implications for sustainable land uses and landscapes. Landsc Urban Plan 157:1–13. https://doi.org/10.1016/j.landurbplan.2016.05.030

Bičík I, Jeleček L, Štěpánek V (2001) Land-use changes and their social driving forces in Czechia in the 19th and 20th centuries. Land Use Policy 18(1):65–73

Ceska pozice (2012) O muži, který vybudoval vlastní přírodní rezervaci. http://ceskapozice.lidovky.cz/o-muzi-ktery-vybudoval-vlastni-prirodni-rezervaci-f8h-/tema.aspx?c=A120602_153235_pozice_68969. Accessed 5 Mar 2018

České Středohoří (2017) Charakteristika oblasti. http://ceskestredohori.ochranaprirody.cz/charakteristika-oblasti/. Accessed 22 Sept 2017

CSOP (2018) Land trusts. http://www.csop.cz/index.php?cis_menu=$cis_menu&m1_id=$m1_id&m2_id=$m2_id&m3_id=m3_id&m4_id=2200&m_id_old=2187. Accessed 20 Mar 2018

CHMI (2017) Územní srážky. http://portal.chmi.cz/historicka-data/pocasi/uzemni-srazky. Accessed 15 Nov 2017

Hanna S, Folke C, Maler KG (eds) (1996) Rights to nature. Island Press, Washington, DC

Hesslerová P, Pokorný J, Semerádová S (2016) The retention ability of the agricultural landscape in the emergency planning zone of the Temelín nuclear power plant and its changes since the 19th century. Land Use Policy 55:13–23. https://doi.org/10.1016/j.landusepol.2016.03.018

Hruška V, Czapiewski K, Kovács Z (2015) Rural economic development in the post-agricultural era: policy recommendations. Studia Obsarow Wiejskich/Rural Stud 39:129–144

Hsieh HF, Shannon SE (2005) Three approaches to qualitative content analysis. Qual Health Res 15(9):1277–1288. https://doi.org/10.1177/1049732305276687

Idnes (2016) Díky tůním nemají lidé suché studny, tvrdí oceněný hospodář. Idnes.cz. https://usti.idnes.cz/rozhovor-mf-dnes-s-danielem-pitkem-d5u-/usti-zpravy.aspx?c=A160726_122555_usti-zpravy_vac2. Accessed 6 Oct 2017

Idnes (2015) Milionář vrací krajině u Milešovky předválečný ráz, i když se mu smějí. Idnes.cz. https://usti.idnes.cz/daniel-pitek-cerncice-milesovka-krajina-fcy-/usti-zpravy.aspx?c=A150520_2163697_usti-zpravy_alh. Accessed 10 Oct 2017

Kaae S, Søndergaard B, Haugbølle LS et al (2010) Development of a qualitative exploratory case study research method to explore sustained delivery of cognitive services. Pharm World Sci 32(1):36–42. https://doi.org/10.1007/s11096-009-9337-5

Kupková L, Bičík I (2016) Landscape transition after the collapse of communism in Czechia. J Maps 12:526–531

Management Plan (2017) Plán péče. http://ceskestredohori.ochranaprirody.cz/cinnost-spravy-chko/plan-pece/. Accessed 16 Nov 2017

Merchant C (2010) George Bird Grinnell's Audubon Society: bridging the gender divide in conservation. Environ Hist 15(1):3–30

Nadace partnerství (2017) Cena Josefa Vavrouška: Daniel Pitek. http://www.cenajosefavavrouska.cz/laureati/daniel-pitek/. Accessed 10 Oct 2017

Neovlivni (2017) Jak se žije ve Středohoří, kde Daniel Pitek vybudoval pevnost proti suchu. Neovlivni.cz. http://neovlivni.cz/jak-se-zije-ve-stredohori-kde-daniel-pitek-vybudoval-pevnost-proti-suchu/. Accessed 11 Nov 2017

Orsillo NP (2008) Agricultural intensification in Communist Czechoslovakia and its impact on the environment. Masters' thesis, Masaryk University, Brno

Pasquini L, Fitzsimons JA, Cowell S, Brandon K, Wescott G (2011) The establishment of large private nature reserves by conservation NGOs: key factors for successful implementation. Oryx 45(3):373–380. https://doi.org/10.1017/S0030605310000876

Pavlínek P, Pickles J (2000) Environmental transitions: transformation and ecological defense in Central and Eastern Europe. Routledge, London, New York

Raška P, Cajz V (2016) Neovolcanic terrain of the České Středohoří Mountains. In: Pánek T, Hradecký J (eds) Landscapes and landforms of the Czech Republic. Springer Verlag, Berlin, Heidelberg, pp 139–152

Respekt (2012) Lovec jelenů na Milešovce. Respekt.cz. https://www.respekt.cz/tydenik/2012/13/lovec-jelenu-na-milesovce. Accessed 14 Oct 2017

Rouillard JJ, Ball T, Heal KV, Reeves AD (2015) Policy implementation of catchment-scale flood risk management: learning from Scotland and England. Environ Sci Policy 50:155–165. https://doi.org/10.1016/j.envsci.2015.02.009

Ruseva TB, Farmer JR, Chancellor C (2016) Networking for conservation: social capital and perceptions of organizational success among land trust boards. Ecol Soc 21(2):540–555. https://doi.org/10.5751/ES-08618-210250

Sklenička P (2016) Classification of farmland ownership fragmentation as a cause of land degradation: a review on typology, consequences, and remedies. Land Use Policy 57:694–701

Slavíková L, Kluvánková-Oravská T, Jílková J (2010) Commentary: bridging theories on environmental governance. Insights from free-market approaches and institutional ecological economics perspectives. Ecol Econ 69:1368–1372. https://doi.org/10.1016/j.ecolecon.2010.02.015

Steinhäußer R, Siebert R, Steinführer A, Hellmich M (2015) National and regional land-use conflicts in Germany from the perspective of stakeholders. Land Use Policy 49:183–194. https://doi.org/10.1016/j.landusepol.2015.08.009

Strategy (2017) Koncepce na ochranu před následky sucha pro území České republiky. Vláda No. 258/2017. Available via DIALOG. http://eagri.cz/public/web/file/545860/Koncepce_ochrany_pred_nasledky_sucha_pro_uzemi_CR.pdf

Tickle A (2000) Regulating environmental space in socialist and post-socialist systems: nature and landscape conservation in the Czech Republic. J Eur Area Stud 8(1):57–78

Wikipedia (2017) Daniel Pitek. https://cs.wikipedia.org/wiki/Daniel_Pitek. Accessed 15 Nov 2017

Yin RK (2003) Case study research: design and methods, 3rd edn. Sage Publication, London

Lenka Slavíková graduated from the University of Economics, Prague (public economics and policy). Currently, she serves as Associate Professor at the Faculty of Social and Economic Studies at JEP University in Ustí nad Labem. Her long-term interest lies in water and biodiversity governance with the focus on Central and Eastern European Countries. She investigates flood risk perception of different actors and financial instruments for flood recovery and mitigation.

Pavel Raška (*1982) graduated from JEP University (history and geography) and Masaryk University (physical geography). He is currently Associate Professor and Head at the Department of Geography, Faculty of Science, JEP University in Ústí nad Labem. His main research interests include flood hazards and risk reduction, in which he explores interdisciplinary perspectives on institutional barriers to FRM and planning.

Chapter 7
Commentary: Mr. Pitek's Land from a Perspective of Managing Hydrological Extremes: Challenges in Upscaling and Transferring Knowledge

Mark E. Wilkinson

Abstract This chapter views the case of Mr. Pitek by looking at the role of NBS in managing hydrological extremes. By viewing the case from a hydrological sciences perspective, we can begin to investigate whether these measures would have an impact in mitigating these extremes (e.g., reducing and delaying a flood peak). Therefore, this chapter gives some suggestions on what information we might need to make this assessment and the challenges and uncertainties in upscaling this approach. Also, the chapter explores if this type of privately funded land management approach is likely to be utilised by other land managers to give a greater density of measures at larger scales.

The Use of Nature-Based Solutions to Manage Hydrological Extremes

Evidence suggests that a warming climate is affecting the timing of river floods in Europe, and projections suggest that climate change will also affect the frequency of floods (IPCC 2012; Bloschl et al. 2017). Similarly, drought risk is also projected to increase over the next decade across most of Europe (Spinoni et al. 2018). Different catchment management approaches can play a role in mitigating these hydrological extremes, for example, using Nature-Based Solutions (NBS), alongside and complementing traditionally engineered approaches. Our catchments also need to provide

M. E. Wilkinson (✉)
James Hutton Institute, Aberdeen, Scotland, UK
e-mail: mark.wilkinson@hutton.ac.uk

T. Hartmann et al. (eds.), *Nature-Based Flood Risk Management on Private Land*,
https://doi.org/10.1007/978-3-030-23842-1_7

other services, such as food production. Intensive agriculture is known to play a role in increasing runoff rates, potentially resulting in flooding and wider water management issues (O'Connell et al. 2007) (this is a common point raised by Mr. Pitek). Many European catchments share an intense focus on farming, a range of environmental issues such as pollution and degraded habitat, highly regulated governance regimes and vulnerability to climate and demographic changes. Hence, agricultural management can become a strategy for managing hydrological extremes but only if the right co-ordination, guidance and compensatory mechanisms are in place. Usually funding is required from public sources to implement catchment-wide NBS over larger scales (e.g., see www.nwrm.eu).

This chapter views the case of Mr. Pitek by looking at the role of NBS in managing hydrological extremes. By viewing the case from a hydrological sciences perspective, we can begin to investigate whether these measures would have an impact in mitigating these extremes (e.g., reducing and delaying a flood peak). Therefore, this chapter gives some suggestions on what information we might need to make this assessment and the challenges and uncertainties in upscaling this approach. Also, the chapter explores if this type of privately funded land management approach is likely to be utilised by other land managers to give a greater density of measures at larger scales.

It is unusual that, as in Mr. Pitek's privately funded approach, a landowner is motivated to invest their own resources to implement NBS measures that additionally lead to a reduction in their farmland's productivity. It is unlikely that many other landowners can spare resources aside from immediate farm business needs; they must instead be motivated by agri-environment subsidies (e.g., funded by the EU Common Agricultural Policy [CAP]). Holstead et al. (2017) found that, in a survey of Scottish farmers, 53% had not installed measures as they considered their land too valuable in its current form, and 38% stated that they did not have measures because of insufficient funding. Other than economics, some other well-established barriers hinder measures at the farm level, as summarised by Holstead et al. (2017):

- Availability of guidance and advice to implement measures.
- Public perception: that a farmer maybe seen as a "slipper farmer".
- Joined-up policy: for example, that installation of new measures may affect the farmer's single farm payment.
- Catchment planning: some landowners are critical of urban planning (e.g., building on floodplains) and see this as contradicting plans to increase the uptake of NFM.
- Traditions: for example, farmers don't favour land rewetting measures relying on previous generations' practice to drain land to increase productivity.

However, these barriers to uptake are not just limited to farmers, and further barriers can be seen in wider institutional bodies (see Waylen et al. 2018).

Nevertheless, the use of pools and wetlands, as presented in this case, is a common option where catchment management schemes have been implemented with a primary aim for FRM (see European Commission 2016). Wetlands and ponds can provide many ecosystem services (Nagabhatla and Metcalfe 2018), whilst temporary storage ponds are starting to become more popular in areas such as North-West

Europe as a means of storing water on agricultural land without impacting productivity too much (Environment Agency 2017) and mitigating the impact of soil erosion by collecting sediments (Evrard et al. 2008). Wetlands have been said to help to alleviate flood risk within catchments (Environment Agency 2017); floodplain wetlands have greater potential for managing floods (Acreman and Holden 2013).

Spatial Complexities and Challenges in Upscaling

Mr. Pitek has been creating pools and wetlands on his land to improve low flows whilst delivering other ecosystem services (e.g., biodiversity enhancement, FRM). If the primary function were to increase groundwater recharge, then the first discussion point would be the optimal placement of these measures in terms of how they connect with the subsurface hydrology. Wetlands may be present in the natural landscape because of impermeable underlying soils or rocks, thus little interaction with the groundwater system occurs. A literature review by Bullock and Acreman (2003) found that, out of 69 literature statements referring to wetland groundwater recharge, 32 assumed without detailed investigation that recharge occurred and 18 suggested no recharge. Finally, six studies found wetlands recharge more, and nine found wetland recharge less than other land types. Therefore, this critical hydrological connectivity of a wetland intending to recharge groundwater systems is suggested by literature to be uncertain, complex and varying from site to site. Wider knowledge is required to understand this spatial complexity, for example, information on geology, soil type and land use.

When considering the role of such pools and wetlands in FRM, the available storage prior to and during a flood event is a key aspect. If a pond is full prior to an event, then it has little available storage to help mitigate the flood runoff (see Quinn et al. 2013). Therefore, one example would be to add a soil bund around the measure to create new "freeboard" storage (i.e., temporary storage). This would allow the pond to store more water during a flood event that could be designed to slowly drain away after the event (e.g., through an outfall or infiltration). This temporary storage approach can be implemented on floodplains and hillslope fields through the creation of bunds (see Environment Agency 2017). This was a common approach applied in the Belford catchment (6 km^2), UK (Wilkinson et al. 2010; European Commission 2016). However, a larger volume of extra storage would be required at larger catchment scales. Metcalfe et al. (2017) suggested that, in a 29 km^2 UK catchment, 168,000 m^3 of extra storage would be required to attenuate peak flow to prevent flooding in a 1.5% Exceedance Probability flood. Mr. Pitek notes that the planning procedure for implementing larger ponds (e.g., 2,000–3,000 m^2) requires greater consent. This suggests that there is a need to review planning procedures to implement larger volume areas (1,000–10,000 m^3) of storage using more natural engineering without falling into the categories of traditionally engineered reservoirs (e.g., >10,000 m^3), where detailed reservoir planning requirements need to be considered, increasing project costs and timescales (see Wilkinson et al. 2013).

As scale increases, the evidence of NBS effectiveness decreases owing to limited empirical studies at larger scales. However, it is at these large scales where policy and planning are more interested in overall combined system performance. In this case, the current plan of the PLA České Středohoří suggests further pools and wetlands are needed to increase the retention capacity of the landscape. Therefore, effective planning and coordination are required to ensure the measures are correctly placed and to minimise any disadvantages at larger planning scales. For example, for FRM, this can involve ensuring measures don't add to the synchronicity of flood peak levels in adjacent tributaries joining a main downstream river but instead try to decouple them (see Lane 2017).

Reflecting on Comparable International Cases

Compared to ponds and wetlands supported by agri-environment and other public funding (e.g., via EU CAP or LIFE funding), those privately funded are likely to be far fewer. However, for different measure types the number of privately funded measures may be greater. For example, a survey in England during the 2013/14 crop year found that 44% of surveyed holdings (and totalling 450,000 ha) had some sort of land in management actions taken from a list of 22 environmental measures that were undertaken without any funding (this includes measures such as grass buffer strips next to a watercourse, fertiliser free permanent pasture, over-wintered stubbles etc.) (DEFRA 2014). Therefore, if the measures generally do not interfere (or show positive coherence) with farming practices, then unpaid measures could be supported by farmers. DEFRA (2014) highlighted that 79% of surveyed farmers responded that protecting soil and water is a primary factor in their land management. However, other studies have noted this motivation alters with different measures, especially those which change land use permanently or impact farm productivity, such as the blocked drains and ponded areas in the Pitek case. Spray et al. (2015) emphasized concerns of farmers in Southern Scotland over potential loss of capital and annual values due to loss of workable land now and in the future. The research suggested that measures involving a reduction in yield or useable land area are not favoured unless a payment mechanism is in place (preferably of annual income, as opposed to one-off compensation).

The notion of privately funded NBS depends on having land available to implement the measures. Therefore, Mr. Pitek is limited not only in regard to his financial resources but also as to when land is available to buy. The approach is not widespread in the case study area, as Mr. Pitek has been unable to motivate his local neighbours. The case stresses a landscape mosaic with many different landowners with measures generally dispersed only on Mr. Pitek's land. How land is owned and managed varies from country to country. In Scotland around 50% of Scotland belongs to 432 owners, including private and conservation charities/trusts (McGuire 2017). These trusts are generally focused on multifunctional landscapes and consider many aspects of land management as opposed to the Pitek case. He noted that conservation trusts are usually focused on one agenda. So, for the case of Scotland, if one of these other large-scale private landowners were to consider investing their own resources into the natural landscape, then the scale of restoration could increase significantly. This is the case with one such landowner who owns over 80,000 ha of Scotland (one of the largest private landowners in Scotland). This owner is creating high-end, nature-based experiences and is interested in the social as well as the ecological needs of the Highlands of Scotland. However, how he and conservation trusts manage the landscape (like Mr. Pitek) may not be "traditional" to some local communities in terms of their management for landscapes to support local livelihoods (going back to the point of traditions mentioned by Holstead et al. 2017).

Globally, in some cases, local communities see water holding measures as an integral part of past traditions and have been reviving historical knowledge and traditions about water management. For example, Rajasthan, India has suffered from extensive drought periods. In the 1980s many of the region's rivers were dry and degraded, where many of the traditional water holding measures were not functioning. Since then over 10,000 Johads (a temporary storage area within a river system designed to hold and infiltrate water) have been constructed by local people to increase recharge into the local aquifer and the river systems are now flowing more regularly. Before the construction of the measures, only ~5% of runoff made it into the aquifer; this has now increased to ~20% (Sisodia 2009). This community-driven initiative has been paramount to the coordination and success of the project (Sisodia 2009).

More empirical evidence is necessary on the effectiveness of catchment approaches for managing hydrological extremes at the catchment scale (Schanze 2017). Therefore, further long-term monitoring of case studies is needed. However, a challenge remains in attaining robust evidence of managed catchments compared to a local control catchment where no management has taken place but where background (e.g., climate) factors can be assessed, and additionally comparing both before and after interventions (see Chap. 6 in Environment Agency 2017). In Mr. Pitek's case, to begin hydrological monitoring after the interventions have been installed may cause challenges in interpreting the evidence.

Concluding Remarks

Currently, the use of catchment-wide NBS in most parts of Northern Europe are focused on flood rather than drought management. However, some catchment planners are now starting to give more consideration to low flow management (e.g., Holt 2018, describes how Slovakia is currently looking at national plans to drought management through NBS). Therefore, the Pitek case offers useful insights into the process of constructing pools and wetlands on the ground (using private funds) to manage low flows and provide wider benefits. If the case were monitored alongside a suitable control area, this could present a useful empirical evidence case study. However, to upscale, researchers would need to gain an understanding how many water holding measures might be needed to influence low flows. The case also highlights the need for central coordination, support and planning for implementing measures over larger scales. The use of decision support tools and opportunity maps could allow for complex information to be used more easily and readily by landowners, implementors and planners (Mackay et al. 2015). For privately funded cases, accessible support mechanisms are needed to allow landowners to access advice on matters such as (a) working in the optimal places, (b) ensuring the measure is designed correctly to deliver its desired ecosystem service, (c) suggesting how measures could be adapted to deliver wider ecosystem services. Mr. Pitek's case also suggests a need to improve the legislative and administration process (e.g., the licencing paperwork) so that it is not seen as a burden that puts people off from implementing measures. In summary, it is vital that these initiatives are supported by offering accessible guidance and acknowledgement given to those who wish to use their private finances in helping to improve the environmental and ecological qualities of our landscapes. For that, it is only fitting that Mr. Pitek has been acknowledged with prestigious awards.

Acknowledgements Open access of this chapter is funded by COST Action No. CA16209 Natural flood retention on private land, LAND4FLOOD (www.land4flood.eu), supported by COST (European Cooperation in Science and Technology).

References

Acreman M, Holden J (2013) How wetlands affect floods. Wetlands 33:773–786

Bloschl G, Hall J, Parajka J, Perdigao RAP, Merz B, Arheimer B et al (2017) Changing climate shifts timing of European floods. Science 357:588–590

Bullock A, Acreman M (2003) The role of wetlands in the hydrological cycle. Hydrol Earth Syst Sci 7:358–389

DEFRA (2014) Campaign for the Farmed Environment (CFE). In: Department for Environment Food and Rural Affairs (ed) Survey of land managed voluntarily in 2013/14 farming year, England

Environment Agency (2017) Working with natural processes—evidence directory. In: Environment Agency for England and Wales (ed) Environment agency, Horizon House, Bristol

European Commission (2016) Natural water retention measures website. http://ec.europa.eu/environment/water/adaptation/ecosystemstorage.htm. Accessed 2016

Evrard O, Vandaele K, Van Wesemael B, Bielders CL (2008) A grassed waterway and earthen dams to control muddy floods from a cultivated catchment of the Belgian loess belt. Geomorphology 100:419–428

Holstead KL, Kenyon W, Rouillard JJ, Hopkins J, Galán-Díaz C (2017) Natural flood management from the farmer's perspective: criteria that affect uptake. J Flood Risk Manag 10(2):205–218

Holt E (2018) Can drought be prevented? Slovakia aims to try. Inter Press Service News Agency, Bratislava

IPCC (2012) Managing the risks of extreme events and disasters to advance climate change adaptation. A special report of working groups I and II of the intergovernmental panel on climate change. Cambridge University Press, Cambridge

Lane SN (2017) Natural flood management. Wiley Interdisc Rev Water 4(3)

Mackay EB, Wilkinson ME, Macleod CJA, Beven K, Percy BJ, Macklin MG et al (2015) Digital catchment observatories: a platform for engagement and knowledge exchange between catchment scientists, policy makers, and local communities. Water Resour Res 51:4815–4822

McGuire T (2017) Who are Scotland's biggest landowners? The Scotsman, Wednesday 11 October 2017

Metcalfe P, Beven K, Hankin B, Lamb R (2017) A modelling framework for evaluation of the hydrological impacts of nature-based approaches to flood risk management, with application to in-channel interventions across a 29-km^2 scale catchment in the United Kingdom. Hydrol Process 31(9):1734–1748

Nagabhatla N, Metcalfe CD (eds) (2018) Multi-functional wetlands; pollution abatement and other ecological services from natural and constructed wetlands. Springer

O'Connell E, Ewen J, O'Donnell G, Quinn P (2007) Is there a link between agricultural land-use management and flooding? Hydrol Earth Syst Sci 11(1):96–107

Quinn PF, O'Donnell GM, Nicholson AR, Wilkinson ME, Owen G, Jonczyk J, et al (2013) Potential use of runoff attenuation features in small rural catchments for flood mitigation. In: Newcastle University (ed) Newcastle upon Tyne, Newcastle

Schanze J (2017) Nature-based solutions in flood risk management—buzzword or innovation? J Flood Risk Manag 10:281–282

Sisodia M (2009) Restoring life and hope to a barren land: 25 years of evolution. Tarun Bharat Sangh, Jaipur

Spinoni J, Vogt JV, Naumann G, Barbosa P, Dosio A (2018) Will drought events become more frequent and severe in Europe? Int J Climatol 38:1718–1736

Spray CJ, Arthur S, Bergmann A, Bell J, Beevers L, Blanc J (2015) Land management for increased flood resilience. In: Crew CRW2012/6 (ed) https://www.crew.ac.uk/publications

Waylen KA, Holstead KL, Colley K, Hopkins J (2018) Challenges to enabling and implementing natural flood management in Scotland. J Flood Risk Manag 11:1078–1089

Wilkinson ME, Holstead KL, Hastings E (2013) Natural flood management in the context of UK reservoir legislation. Centre of Expertise for Waters, Aberdeen

Wilkinson ME, Quinn PF, Welton P (2010) Runoff management during the September 2008 floods in the Belford catchment, Northumberland. J Flood Risk Manag 3(4):285–295

Mark Wilkinson is a senior research scientist in catchment hydrology at the James Hutton Institute. The focus of his work is designing and investigating the impact of NBS with the main aim to reduce and attenuate flood peaks. He is also interested in the multiple benefits associated with these measures.

Chapter 8
Commentary: A Spatial Planning Perspective on Privately Funded Natural Water Retention Measures

Lukas Löschner

This contribution puts privately funded natural water retention measures (NWRM) in a spatial planning perspective. It comments on the case presented by Slavíková and Raška (this volume), which investigates the role of a forest engineer and farmer (Mr. Pitek) who acquired and restored degraded agricultural areas by improving the water retention capacities and developing more sustainable farming practices. The case is exceptional because it illustrates that private initiatives may actually support, and not only undermine, public policy goals concerning the enhancement of biodiversity and landscape ecology. It thus stands in contrast to related studies, which typically explore the efforts of public authorities and their respective policy instruments to implement NWRM. Such studies often identify private landowners as a key target group of policy interventions (due to their private rights to property and land use), but focus on the policy means by which to incentivise NWRM or mobilise land to facilitate their realisation. Instead of placing private landowners at the receiving end of environmental policies, this study provides a bottom-up landowner perspective. It investigates the personal motivations for restoring the ecological quality of landscapes and provides valuable insights into the (mal)functioning of specific policy instruments (e.g., agricultural subsidies) and the bureaucratic procedures involved in landscape restoration.

Although Mr. Pitek's initiative to restore water retention ponds is principally in line with the overall aims of environmental policy in the region, he makes a point "to do what he considers to be the right thing without subordinance to public authorities and/or subsidy schemes" (Slavíková and Raška, this volume). He has purposely chosen to build a series of small retention pools "to avoid complicated permitting processes related to changes in land use designation". In fact, Mr. Pitek's plans to install larger ponds remain to be realized due to lengthy authorization processes and

L. Löschner (✉)
Department of Landscape, Spatial and Infrastructure Sciences, Institute of Spatial Planning, Environmental Planning and Land Rearrangement (IRUB), University of Natural Resources and Life Sciences Vienna (BOKU), Vienna, Austria
e-mail: lukas.loeschner@boku.ac.at

T. Hartmann et al. (eds.), *Nature-Based Flood Risk Management on Private Land*,
https://doi.org/10.1007/978-3-030-23842-1_8

he "seems to be frustrated with this bureaucracy". He is moreover highly sceptical of agricultural subsidies, which he criticizes for "sending perverse incentives to land management and agricultural production"; "instead of supporting measures, such as those aimed at water retention and biodiversity (…) subsidies [are provided] to intensive production of crops for biofuels".

Mr. Pitek's satisfaction "with the possibility to restore pools on his own—cheaply and fast" illustrates that the prevailing set of regulatory and financial instruments in Czech environmental policy is apparently not suited to foster NWRM on privately owned land. While the effectiveness of agricultural subsidies and regulations is a critical issue in this regard, the aim of this contribution is to place the above case in a spatial planning perspective. In this vein, it provides a reflection on the opportunities and limitations of spatial and land policy instruments to implement NWRM and discusses the contribution of this study for related research in environmental policy.

Securing Land Resources for NWRM with Spatial Planning Instruments

Generally speaking, spatial planning refers to "the geographical expression, implementation, and coordination of public policy across sectors and scales. (…) By creating spatial plans [it] expresses where and in what form policy will unfold, coordinates and aligns initiatives to avoid duplication of effort or divergent policies being adopted" (Castree et al. 2013, p. 485). Regulatory instruments of spatial planning—for example, regional spatial plans, local zoning plans or building schemes—can thus play an important role to legally secure the necessary land resources for NWRM. Regional spatial plans, for instance, may be used to designate large-scale areas for flood retention or flood runoff (see Löschner et al., this volume). Similarly, regional spatial plans may be used to preserve wetlands and natural floodplain, not only as a means to attenuate runoff attenuation and propagate flood waves, but mainly to preserve and enhance the biodiversity of riverine ecosystems. Local zoning plans are also effective formal planning instruments to allocate land for green and blue infrastructure, while building schemes may, for instance, be used to define small-scale rainwater detention basins on the plot-level and thus contribute towards better municipal rainwater management in the face of climate-induced increases in pluvial floods.

These (formal) spatial planning instruments are, however, overwhelmingly limited to settlement areas and building land. In general, spatial planning has little leverage on agricultural or forestry land. Apart from assessing the compatibility of land use plans, spatial planning does not dispose of the necessary regulatory means to infringe on agricultural land and is therefore not able to play a significant role in facilitating NWRM in these areas.

Mobilising Land Resources for NWRM Through Land Consolidation

Given these limitations, it may be useful to also consider instruments of land policy (as a closely related discipline) and take a look in particular at the instrument of land consolidation and its possibilities to realize public aims on privately-owned agricultural and forest land. Land consolidation is an instrument of land management to enlarge, redesign, rearrange and improve agricultural and forest land. As indicated, for instance, in the Austrian federal Framework Land Consolidation Act, the objective of land consolidation is to improve and rearrange the conditions of agricultural cultivation on behalf of an efficient and environmentally friendly agriculture by consolidation and development of agricultural and forest land (Mansberger and Seher 2017). Land consolidation schemes generally demand a bottom-up initiative and a common interest on the side of the landowner(s) to rearrange parcels of agricultural land for a certain benefit. A land consolidation scheme may thus be started if disadvantages in farming structure (land fragmentation of unfavourable plot size) shall be eliminated.

Although the core task of land consolidation is the improvement of farming structures, in Austria the instrument is also applied to obtain the necessary land resources for the public benefit, such as the development of traffic infrastructure. In this vein, multifunctional land consolidation schemes are also increasingly implemented to address non-agricultural, land-related issues, such as acquiring land for flood retention and flood protection measures, as well as for river restoration and nature conservation. Land acquisition for flood plains, considering both public interests and landowners' expectations can be realised by means of land consolidation. The substantial strength of this procedure lies within the mobility of land, as parcels may be purchased anywhere within the consolidation area and allocated where needed. This allows, for instance, for the creation of coherent areas for flood runoff and flood retention, which would normally not be possible through land acquisition, given the oftentimes large number of riparian landowners. Nevertheless, multifunctional land consolidation is strongly dependent on the availability of land required by planning authorities. In the future, restrictions for this type of land consolidation can be expected due to increasing agricultural land use demands as a result of higher prices for agricultural products and the production of renewables (Seher 2015).

Further Questions and Lessons Learned

The above spatial planning and land policy instruments are of course only applicable if a need to provide (private) land for a specific environmental policy goal exists, such as natural water or flood retention. In the case of Mr. Pitek, however, agricultural practices actually align with and in many ways support public aims to increase biodiversity and water retention capacities in degraded agricultural areas.

The principal question concerning this case therefore is not how we may secure or mobilise the necessary land resources to better fulfil a public policy aim, but rather how comparable initiatives by private landowners may be promoted.

In this regard, Slavíková and Raška rightly stress that the current agricultural subsidy scheme is too fixated on agricultural production and is therefore in need of reconsideration to also provide incentives for multifunctional land uses and more locally attuned cultivation methods. They, however, also highlight the central role of land ownership—in particular given Czechia's rather recent transition to a market economy—to effectively implement NWRM: to Mr. Pitek, "the only way forward (…) was to become a landowner and to undertake changes on his own property". Owning the land allows him to get things done "fast and cheap (although non-participatory and to some extent non-expert)". This puts planning authorities into a bit of a dilemma because it shows the difficulties of involving such hidden champions in a more coordinated environmental policy scheme. Although Mr. Pitek's agricultural approach has drawn increasing scholarly attention and shows apparent benefits in terms of an observed increase in biodiversity as well as agricultural yield, "no permanent monitoring" to assess the actual hydrological effects and ancillary benefits of his actions has been put into place.

With the aim of bringing private initiatives and public authorities together, a big challenge therefore appears to be to build a better understanding of the different rationales but maybe also how to build mutual trust. Slavíková and Raška mention that efforts have been made to promote "participatory governance schemes" in this field. Especially in a former Communist country with a weak tradition in participatory planning approaches, it would be interesting to learn more about these governance schemes and processes: What are their specific aims? How are they organised? Who is involved and why is it so challenging to get people "on board"? Which role could people like Mr. Pitek play in such processes? Delving deeper into these issue could provide further insights into how private initiatives and public aims in environmental policy can be more clearly aligned.

Overall, the account of Mr. Pitek, however, is highly valuable because it gives a rare stakeholder-centred perspective on environmental policy. It shows that research can benefit from zooming in on a landowner perspective to obtain complementary insights to the often policy-centred research. Apart from learning about the farmers' intrinsic motivations, the case also provides interesting insights into different policy instruments. It thus highlights that—when investigating policy instruments (be they in spatial planning, land policy or agriculture)—it is critical to not just look at "how intentions of policy are translated into operational activities" (de Bruijn and Hufen 1998, p. 12) but to also account for those targeted by the policy instruments. Such a perspective may help understand in how far subsidies, regulations or other policy instruments are able to achieve their desired effect (e.g., by delivering changes in land use and fostering NWRM). At the same time it is crucial to keep in mind that policy instruments are generally not simply functional in design, but instrument choice actually depends on a number of different factors, including the institutional context, interest groups etc. (Linder and Peters 1989). Bringing the different perspectives—stakeholder-policy and substantive-process—in the research on policy instruments together can contribute to developing a more encompassing understanding of environmental policies.

Acknowledgements Open access of this chapter is funded by COST Action No. CA16209 Natural flood retention on private land, LAND4FLOOD (www.land4flood.eu), supported by COST (European Cooperation in Science and Technology).

References

Castree N, Kitchin R, Rogers A (2013) A dictionary of human geography. Oxford University Press, Oxford

de Bruijn H, Hufen H (1998) The traditional approach to policy instruments. In: Peters BG, van Nispen FKM (eds) Public policy instruments: evaluating the tools of public administration. Edward Elgar Publishing, Cheltenham, pp 11–32

Linder SH, Peters BG (1989) Instruments of government: perceptions and contexts. J Publ Policy 9(1):35–58. https://doi.org/10.1017/S0143814X00007960

Mansberger R, Seher W (2017) Land administration and land consolidation as part of Austrian land management. EU Agrarian Law 6(2):68–76. https://doi.org/10.1515/eual-2017-0010

Seher W (2015) Potenziale der Grundzusammenlegung als Instrument des Flächenmanagements in ländlichen Räumen Österreichs. Zeitschrift für Verwaltung (ZfV) 6:365–372. https://doi.org/10.12902/zfv-0082-2015

Dr. Lukas Löschner has an academic background in Political Science and Landscape Planning and holds a Ph.D. in Spatial Planning. His main field of research is natural hazard risk management, which he explores in a combination of political science and planning approaches. He currently conducts research on the following topics: spatial adaptation to flooding, policy coordination in FRM, flood risk governance, and flood storage compensation.

Part IV
Floodplain Development in Flanders

Chapter 9
Swapping Development Rights in Swampy Land: Strategic Instruments to Prevent Floodplain Development in Flanders

Ann Crabbé and Tom Coppens

Creating natural flooding areas to give way to water requires space. In practice, it is often difficult to deal with historically designated development rights, particularly in densely populated areas. Development rights are often seen as a part of land property rights of a land title and represent a certain financial value. Our contribution aims to describe how Flanders, the northern and Dutch speaking region of Belgium, struggles with (re)allocating development rights in flood-prone areas, highlighting that regular and innovative policy instruments significantly shape the behavioural responses of different actors in flood plains. In our article, we discern instruments for reallocation based on market approaches, government approaches and community approaches. From the Flemish context, we learn that (re)allocating development rights through government-based initiatives with financial compensations is still considered the most feasible approach, although the costs for the government may escalate. Even though Flanders is working with innovative instruments like trading development rights and exchange of development rights, they are expected not to fulfil the promise of becoming a true alternative for the central government-led initiatives. The main reason for that is that incentives to trade development rights or pool land are lacking as there is no real advantage for landowners nor municipal governments in participating in trade or pooling.

A. Crabbé (✉)
Faculty of Social Sciences, Centre of Research on Environmental and Social Change (CRESC), University of Antwerp, Antwerp, Belgium
e-mail: ann.crabbe@uantwerpen.be

T. Coppens
Faculty of Design Sciences, Research Group for Urban Development, University of Antwerp, Antwerp, Belgium
e-mail: tom.coppens@uantwerpen.be

T. Hartmann et al. (eds.), *Nature-Based Flood Risk Management on Private Land*,
https://doi.org/10.1007/978-3-030-23842-1_9

Introduction

Climate change and increasing land coverage have an impact on flooding risks, even in areas that historically didn't have problems with water (Hellmann and de Moel 2014; Wheater and Evans 2009). In the evolution from technocratic "vertical" flooding policies to a socio-ecologic "horizontal" approach, most governments aim to restrict urbanisation of risk areas, in order to make "space for rivers" (Warner et al. 2012). The "space for rivers" thereby reflects a paradigm shift from a technological approach oriented toward containing and restraining water flows with the help of dikes and artificial basins versus a nature-based approach with loosely controlled natural flooding areas.

Creating natural flooding areas to give way to water requires space. In practice, it is often difficult to deal with historically designated development rights in these areas, in particular in densely populated areas. Development rights are often seen as a part of land property rights of a land title and represent a certain financial value (Nelson et al. 2013). Changing the development rights therefore results in a financial loss for the property owners. Property owners in flooding areas tend to oppose government initiatives to restrict new developments that often lead to a gap between water policies' goals and their effective implementation.

The most common instrument to (re)allocate development rights is through zoning or rezoning. However, this instrument is often not considered attractive, due to long procedures and expensive financial compensations. Next to zoning, alternative (innovative) instruments are being developed: (a) the system of tradable land development rights, which is a market-based instrument and (b) the system of exchange of development rights and land pooling (see box in Fig. 9.1).

Our contribution aims to describe how Flanders, the northern and Dutch-speaking region of Belgium, struggles with (re)allocating development rights in flood-prone areas, highlighting that regular and innovative policy instruments significantly shape the behavioural responses of different actors in flood plains. In the next section, we briefly describe flood risks in Flanders and some core characteristics of land use policy in Flanders. We introduce the idea of designating "signal areas" as a means to give policy priority to these areas where the planned land use conflicts with the potentially high flood risk damage. We discuss that changing development rights via (government-initiated) spatial implementation plans does not seem to be a very effective and time-efficient solution. This explains why the Flemish government decided to develop complementary and alternative instruments: trading development rights and mandatory land readjustment. We state that these innovative approaches are not (yet) a true alternative for a central government-led approach as the necessary incentives to trade development rights or pool land are structurally lacking.

Box 1: Three types of solutions for land management

Governance problems such as water and land use management can typically be approached by three types of coordination mechanisms: markets, hierarchies and networks (Thompson, 1991) . In the table below, we discern three types of solutions for the governance land management: a government-based, a market-based and a community-based solution. (a) The government-based solution is the most common, with a hierarchical government that intervenes with the idea that its intervention can influence or stimulate welfare for all. Classical land management instruments in a government-based approach are zoning, expropriation and (right of) pre-emption. (b) Market-based solutions and community-based solutions are much less practiced. In the market-based solution, land is considered an economic good with demand for and supply of land from societal/economic actors, that transact property rights. The government only intervenes in a facilitating role, not in an authoritative function. Transferable development rights are a typical land management instrument in a market-based approach (Nelson et al., 2013; Ward, 2013). (b) In a community-based approach land is considered part of the commons, for which collective action is needed. In that approach, citizens need negotiation and mutual coordination to settle land issues and develop institutions that govern common affairs (Ostrom, 1990). A typical land management instrument that fits a community-based approach is (voluntary or mandatory) land adjustment (van der Krabben & Needham, 2008)

Type of solution for land management	Government based solution	Market based solution	Community based solution
Allocation mechanism	Central planning	"Invisible hand"	Negotiation and mutual coordination
Underlying theories	Interventionist welfare economics theories	Transaction costs and property right theories	Commons, theories on collective action
Typical land management instruments	Zoning, expropriation, right of pre-emption	Transferable development rights	Voluntary land re-adjustments Mandatory land re-adjustment

Fig. 9.1 Three types of solutions for land management

Flood Risks and Land Use Policy in Flanders

The Risk of Floods in Flanders

Flood risk control poses an important challenge in Flanders, the low-lying part of Belgium. Geographically, it consists in a coastal basin plain in the north-west and a central plain, crossed by the river basin of the river Scheldt, the Yzer and the Meuse and their tributaries. With an average density of about 485 inhabitants per km^2, the region has a high level of urbanization. Moreover, as the historical urbanisation mainly took place along the rivers, the river valleys tend to be the most populated areas, making them especially vulnerable to flood risks.

According to the OFDA/CRED international disaster database, the frequency of problematic floods has been increasing over the last years in Belgium (see www.emdat.be). River flooding occurs due to a combination of heavy rainfall and long periods of rain. Also sea storms can pose risks for cities along the estuaries of the Scheldt and Yzer. According to the Flanders climate report, 7.5% of the total surface is now exposed to an increased risk of flooding, affecting 220 000 inhabitants. The same report estimates the average yearly costs of flooding damage at about EUR 50 million.

Current models predict an increase of flooding in the future. Although the exact impact of climate change on flood risk is still unclear (Brouwers et al. 2009), it is expected that changing land use patterns and soil sealing will result in an increase in peak loads of the river system. The total of sealed soil increased from 4 to 5% in 1976 to 12.9% in 2012, whereas the total share of urbanized land now approaches one third of the total surface (Departement Omgeving 2017). According to the Flemish administration, about 6 ha per day are converted to urban use. The high levels of urban sprawl in Flanders further present a particular challenge to flood control (De Decker 2011; Verbeek et al. 2014), with ribbon development and scattered, low-density development.

Land Use Policy in Flanders

In the federal state of Belgium, the competencies on spatial planning and environment are situated at the regional level. The current planning framework in Flanders is determined by the 1962 Belgian act on spatial planning, the 1997 structure plan, the 1999 decree and the 2009 Codex on spatial planning.

The Belgian 1962 act introduced a comprehensive set of hierarchical zoning plans on different scale levels. The plans on the regional level or the *gewestplannen*, which were developed and approved in the 1970s and the early 1980s, have been seminal for the Belgian planning system. They provided detailed zoning prescriptions for the whole territory that are still in place today (for an example, see Fig. 9.2). Because of an overestimation of the growth during the making of the *gewestplannen*, but also

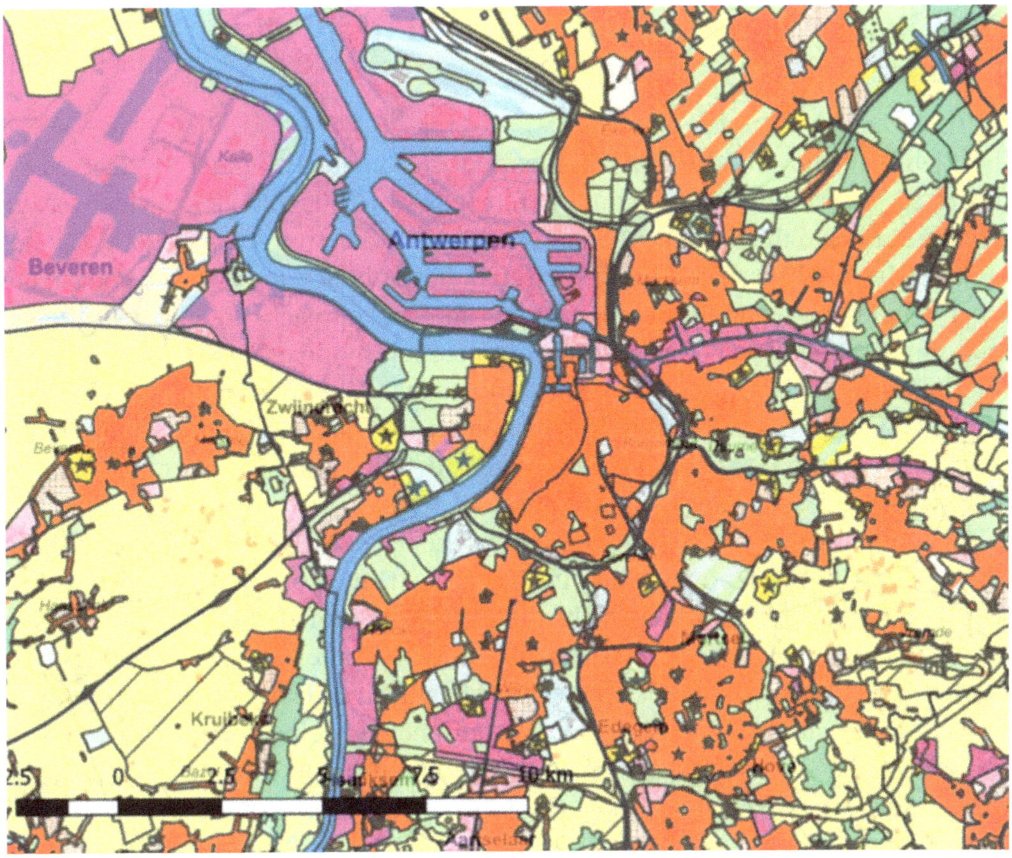

Fig. 9.2 The “gewestplan”: land covering comprehensive zoning instrument. *Source* Extract of the regional zoning plan for Antwerp and surrounding municipalities; made in QGIS, based on the regional dataset of the ‘gewestplan’

political lobbying from local governments resulting in an excess of supply in housing areas, many areas designated for development are still left unbuilt. The Flemish government estimates that about 41 000 ha of designated land in the *gewestplannen* are still available for housing and mixed use developments, providing a large “juridical stock” of buildable land. A sizable share of this stock is located in water-prone areas.

The majority of this land belongs to private landowners who have development rights that are not restricted in time. Although one needs a building permit to build, in practice building permits in designated housing areas cannot just be refused without compensation. Moreover, the zoning of the *gewestplannen* has had a dramatic impact on land prices that fuelled land speculation. Whereas areas designated as “housing areas” have an average price of about EUR 180/m^2 in Flanders (2014), the value of land zoned as agricultural land swings between EUR 4–5/m^2. This financial aspect makes changing the planned land use in Flanders particularly challenging.

Flood Risks Policy (Instruments) Flood Risk Policy in Flanders

The Flemish policy on flood risk prevention is regulated by the 2003 Decree on Integrated Water Policy and the policy framework of 2005 and 2013, which assign implementation to the European Water Framework Directive of 2000. Severe floodings in 2010 and 2011 marked an increased policy attention to flood risk control but also a "window of opportunity" to implement another approach. Traditional technical flood protection policies oriented toward protection by dikes, retention basins and water pumping installations have been complemented by prevention and NBS with natural flooding areas and "natural" retention areas. This more ecological approach affects land use policy and in particular areas that are designated for housing and industry in flood risk areas.

The government has established a policy framework in 2013 to deal with new developments in flood-prone areas. In this context, the concept of "signal areas" (*signaalgebieden*) has been created: these are areas designated for industry or housing and with a potential impact on the water system. If after a further analysis it is concluded that flood damage risk increases when the area is developed according to the zoning designation, then the Flemish government concludes that this area needs a follow-up phase. In this follow-up phase, the Flemish government determines the development perspective for the area. Three development options are possible:

- Option A: the current designation is compatible with the water retention function, and building is possible with limited restrictions;
- Option B: the current designation has a moderate impact on flood risks, and building is possible with severe restrictions. Option B includes water-proof construction techniques and adaptive buildings;
- Option C: the current designation has a substantial impact on flood risks and building is prohibited.

Based on intensive discussions between the spatial planning department of the Flemish Government and other departments within the Coordination Commission on Integrated Water Policy, 3338 ha of land have now been labelled as signal area. About 2473 ha fall under development option C and need to be rezoned.

Instruments for Land Use Management in Flood-Prone Areas

The policy framework on the signal areas proposes a number of policy instruments to restrict and prohibit building in flood-prone areas. In the option A and option B areas, a number of instruments intend to prevent or mitigate the impact of new construction.

- Building permits in areas with a flood risk need a water test. Water managers advise here regarding the following: (a) the potential impact of the building on the water system and (b) measures to mitigate or compensate these impacts.
- The instrument of information obligation includes that potential buyers or renters in risk-prone areas have to be informed explicitly regarding the risks of flooding when purchasing real estate.
- The governments also provides guidelines for adaptive building in water-prone areas; these are building construction techniques that reduce the impact on the water system and are able to cope with potential flooding.

For option C areas, zoning designations have to be changed, and more drastic instruments are needed. Zoning designations can be changed by spatial implementation plans or *ruimtelijke uitvoeringsplannen*. The Codex on spatial planning regulates the procedure, which foresees a public inquiry, and multiple rounds of advisements from both government administrations and advisory commissions. The procedure typically takes 2–3 years; implementation plans for more complex projects can even last 10 years. Implementation plans can be initiated by the municipal, provincial and regional level as long as they fit within their strategic structure plan. Implementation plans are also subject to environmental impact studies, which are regulated by separate environmental legalization.

Moreover, the 1962 Act and the later Codex protected landowners from value destruction as a result of rezoning in implementation plans with the instrument of plan compensation (*planschade*). Landowners that are faced with rezoning are compensated for 80% of the acquired value at current prices. Moreover, when option C is decided for a flood-prone area, it must be determined which level of the government (municipal, provincial or regional) has to take the initiative to start the procedure to make a spatial implementation plan or *ruimtelijk uitvoeringsplan*. Important to mention is that plan compensations have to be paid by the government level that takes the initiative. Local governments can rely on regional subsidies, but only for up to 60% of the costs for the plan compensations.

Because of the complexity of the instrument and the financial implications, governments are mostly reluctant to use implementation plans. An evaluation of the progress of redesignation in 83 signal areas found that for three quarters of the signal areas that need planning redesignation, procedures to develop spatial implementation plans had not been initiated yet. The Flemish government initiated 16 regional spatial implementation plans on signal areas; some provincial governments have also made spatial implementation plans on water-prone areas; but for municipal governments, it proved very difficult to redesignate water-prone areas via spatial implementation plans. The municipal governments indicate as a main reason a lack of clarity about the financial consequences and the fear for having to pay (large amounts of) financial compensations due to the development restrictions for landowners. While municipal governments are hesitant to initiate spatial implementation plan procedures, water-prone area are being developed, leading to high(er) risks of flood damage in these areas.

Flood Prone Open Space Areas as a Bypass

In order to safeguard water-prone areas from further development, the Flemish government recently introduced the instrument of "flood prone open area spaces". With this instrument, the Flemish government has made it possible to change the current designation of signal areas to open space area. The "open space" designation leaves open which particular designation the area will receive (nature, agriculture, soft recreation…), but firmly confirms that the area cannot be developed. Some small developments will still be allowed, such as building little shelters as bus stops or building a functional road through the area, but high value development with residential homes and industry will not be allowed. Which areas will receive a flood-prone open space area designation will not be decided upon in a spatial implementation plan but in a circular note of the Flemish government.

Giving signal areas the planning designation of a flood-prone open space area via a circular note of the Flemish government has many advantages: (1) the responsibility to change planning designation is shifted from other governments towards the Flemish government, which releases more local governments from making unpopular decisions in complex spatial implementation plan procedures; (2) it adds speed to the process of taking away development rights in signal areas; (3) it creates more clarity and legal security for landowners who own land that is labelled as signal area; and (4) it enables a more centralized and uniform financial compensation policy for landowners who lose development rights.

The Flemish government decides which signal areas will receive a flood-prone open space area designation based on a suggestion by the Coordination Commission on Integrated Water Policy, in which not only the spatial department of the Flemish government but also provinces, municipalities and many other governmental agencies are represented. The decision on which planning redesignation will be assigned—by means of spatial implementation plans or by means of a circular note on water prone open space areas—will be made pragmatically. If procedures for spatial implementation plans have already been initiated, then signal areas will get a planning redesignation by means of a spatial implementation plan. If not, a circular note will be the alternative.

The Biggest Challenge for Implementing the Flood Prone Open Space Area Policy

The 2013 policy framework located the financial responsibility for compensating landowners in flood-prone areas at the government level that takes the initiative to make an implementation plan. In the new regulation, the minister for spatial planning has expressed that all compensations would be financed by the regional level. Moreover, the landowner will receive a financial compensation—not based on indexation of the purchase price but on the actual market value of the land. This means that financial compensation for losing development rights risks will become much more expensive for the Flemish government because purchase prices of land (even though indexated) are much lower than the actual market value of land. The spatial planning department therefore is urged to build more expertise in determining the actual market value of land by trying to systematically estimate the market value price of flood-prone signal area land. In these exercises, the spatial planning department starts off with the idea that market prices of signal areas are in practice lower because buyers anticipate that governments will not allow building on that land due to the signal area status.

The high cost of financially compensating the loss of development rights, together with the lack of experience and expertise on determining actual market prices of signal area land, are—at this moment—important potential drawbacks in implementing the flood-prone open space area policy of the Flemish government.

Are There Alternatives to Implementation Plans?

Because of the bureaucratic complexity and the financial impact of implementation plans, the Flemish government has decided to develop complementary and alternative instruments.

Trading Development Rights

A new decree is underway with instruments to create a "market place" to trade development rights. The instrument of transferable development rights differentiates between sending areas and receiving areas; owners of flood-prone areas (sending areas) are compensated by owners buying development rights in the receiving areas.

In Flanders, municipal governments in particular are interested in the transferable development rights instrument. When a market place is created (when the transferable development rights-instrument is implemented), municipalities are relieved from the burden of financially compensating landowners because owners of areas receiving development rights financially compensate the owners of the sending areas.

The idea of transferring development rights on a "market place" is not really convincing for many. One has the illusion that the transferable development rights instrument would help to limit or take away development rights on land that should ideally not be built on, but it probably will not. Neither the owners of sending land nor the owners of receiving land seem to be very enthusiastic. (a) Owners of the sending area are not very keen on entering a market where the price of their land and/or development rights depend on the demand of others. They feel much more comfortable with the option that the government would fully financially compensate them for the loss of development rights and do not want to risk a volatile "bidding" among owners over receiving areas. (b) Owners of receiving areas do not have a proper incentive to pay for extra development rights on their land (e.g., building higher apartments), because the development rights on their (receiving) land already are rather permissive. Or, to put it differently, they are not pushed by "scarcity" that prevented them from paying for extra developments rights. They do not want to pay for something that is already allowed anyway.

Furthermore, scientific research concluded that the instrument of transferable development rights is not prohibiting open space to be built on (Ruimte Vlaanderen 2016; Van den Nieuwenhof 2016). Van den Nieuwenhof found that in order to withdraw the development rights of one piece of land, you have to attribute seven development rights in receiving areas in order to make the transfer of development rights financially interesting for the owners. Again, the main reason for that is that regulations on developments were and are quite permissive in Flanders. For example, many apartment buildings have been built in Flanders recently. This has caused an oversupply of apartments in the real property market. Developers are not interested in building higher apartment blocks, as they cannot find buyers. This makes attracting extra development rights not very interesting for them.

Mandatory Land Readjustment, Including Swapping Designation Zones

Land readjustment has been a common practice in many countries for decades (van der Krabben and Needham 2008). In Flanders, its purpose until now has been dominantly to pool fragmented properties and to redistribute property rights more efficiently. Typical applications are in agricultural and nature zones where parcels of land are swapped between the owners. The government delineates a district in which land reparcelling is possible, and, under the coordination of a government (land) agency, property owners are invited to collaborate and bring in their land as investment capital. In return, the agency commits to giving each owner a land site of at least equal value in the vicinity of the original site upon the completion of the land reparcelling project.

In case of mandatory land readjustment, the government puts land readjustment on the agenda to create a solution for problems with the planned land use and its associated development rights. There are now a few emerging cases using the instrument to relocate development rights in flood-prone areas. In these cases, the government is eager to swap planning designations between parcels/landowners, in order to prevent flood-prone areas from being developed (further). In these cases, the landowners remain landowners (no expropriation) but are financially compensated for the loss of development rights on their land by swapping planning designations between parcels/landowners, for example, swapping building rights for rights to use the land as agricultural land.

The difference between the classic land reparcelling and the new mandatory land readjustment is that in classic land reparcelling it is about land with "soft" planning designations (such as agriculture and nature) while the mandatory land readjustment brings about an exchange of designations between land with soft designations and hard designations (e.g., residential area or industry zone) where development rights should be revoked.

The idea of mandatory land readjustment provokes the interest of municipal governments, mainly because of financial reasons. In contrast to paying financial compensations in case of a spatial implementation plan, municipalities are not obliged to compensate landowners that lose development rights. When designations are exchanged between parcels/landowners, owners are paid in real values.

In practice, however, the mandatory land readjustment is not expected to be a success. Many reasons can explain this.

(a) If municipal governments approach the Flemish land agency to swap development rights between owners, they often lack a vision for the entire district for land readjustment, while this is a fundamental condition for the Flemish land agency to start a project. If municipal governments have the idea to install a flood retention area on the flood-prone land, than they are obligated to actually implement a flood retention zone in the area. If they want a land readjustment, then the land reparcelling should actually be done within the project. In summary, in contrast to the "flood-prone open space area"-instrument, where the exact designation of the open space can be determined later, mandatory land readjustments with designation swapping implies immediate choices on the final designation of the land and immediate implementation measures. These obligations are expected to keep municipal governments from actually starting up these types of projects.
(b) Landowners are often reluctant to swap designations as they prefer to be financially compensated by the government for losing development rights instead of being compensated in real value. The land readjustment then leads to a spatial implementation plan in which development rights are taken away from parcels, but landowners will ask for extra favours, for example, extra development rights on other parcels or permission to build pale dwellings on their land.
(c) One cannot easily find parcels to match. In Flanders with its strong urbanization, where can one find yet another zone where one could redesignate land with a soft planning designation to land with a hard planning designation?

In Summary: Market- and Commons-Led Initiatives Are not yet a True Alternative for a Central Government-Led Approach

From the Flemish context we learn that (re)allocating development rights through central government-led initiatives with financial compensations is still considered the best approach, because it is the fastest way to avoid (further) development of flood-prone areas and because it is an appealing scenario for municipal governments and landowners. A big drawback to these central government-led initiatives is that committing to these financial compensation carries the risk of becoming too great a financial burden, in particular when the government financially compensates at the rate of the actual market value of the land.

Even though Flanders is working with innovative instruments like trading development rights and exchange of development rights, they are not expected to fulfil the promise of becoming a true alternative for the central government-led initiatives. The main reason for that is that incentives to trade development rights or pool land are lacking since landowners do not see the advantage of participating in trade or pooling.

Acknowledgements Open access of this chapter is funded by COST Action No. CA16209 Natural flood retention on private land, LAND4FLOOD (www.land4flood.eu), supported by COST (European Cooperation in Science and Technology).

References

Brouwers J, Peeters B, Willems P, Deckers P, De Maeyer PH, Vanneuville W (2009) Klimaatverandering en waterhuishouding. In: Van Steertegem M (ed) Milieuverkenning 2030: milieurapport Vlaanderen, Vlaamse Milieumaatschappij, pp 283–304

De Decker P (2011) Understanding housing sprawl: the case of Flanders, Belgium. Environ Plann A 43(7):1634–1654. https://doi.org/10.1068/a43242

Departement Omgeving (2017) Ruimterapport Vlaanderen, Open Ruimte. Brussel

Hellmann FA, de Moel H (2014) Future land use patterns in European river basins: scenario trends in urbanization, agriculture and land use. In: Risk-informed management of European river basins. Springer, Berlin, Heidelberg, pp 209–222. https://doi.org/10.1007/978-3-642-38598-8_7

Nelson AC, Pruetz R, Woodruff D (2013) The TDR handbook: designing and implementing transfer of development rights programs. Island Press, Washington, D.C.

Ostrom E (1990) Governing the commons: the evolution of institutions for collective action, 1st edn. Cambridge University Press, Cambridge/New York

Ruimte Vlaanderen (2016) Adviesnota Verhandelbare Ontwikkelingsrechten. Available via DIALOG. https://www.ruimtevlaanderen.be/Portals/108/Adviesnota_verhandelbare_ontwikkelingsrechten_2016.pdf

Thompson G (1991) Markets, hierarchies and networks: the coordination of social life. SAGE Publishing, London

Van den Nieuwenhof L (2016) Toepasbaarheid van verhandelbare ontwikkelingsrechten in Vlaanderen: is het concept verhandelbare ontwikkelingsrechten toepasbaar in gemeentelijk beleid in Vlaanderen? Master's thesis, University of Antwerp

van der Krabben E, Needham B (2008) Land readjustment for value capturing. A new planning tool for urban redevelopment. Town Plann Rev 79(6):651–672

Verbeek T, Boussauw K, Pisman A (2014) Presence and trends of linear sprawl: explaining ribbon development in the north of Belgium. Landscape Urban Plann 128:48–59. https://doi.org/10.1016/j.landurbplan.2014.04.022

Ward P (2013) On the use of tradable development rights for reducing flood risk. Land Use Policy 31:576–583. https://doi.org/10.1016/j.landusepol.2012.09.004

Warner JF, van Buuren A, Edelenbos J (2012) Making space for the river. IWA Publishing, London

Wheater H, Evans E (2009) Land use, water management and future flood risk. Land Use Policy 26:251–264. https://doi.org/10.1016/j.landusepol.2009.08.019

Ann Crabbé is guest professor and senior researcher at the Research Centre for Environmental and Social Change at the Sociology department of the University of Antwerp (Belgium). Her research interests include institutional stability and dynamics in the governance of flood risks. After her Ph.D. on institutionalizing the river basin approach in Flanders, she was involved in several regional and European projects on water quality policies and flood risk governance.

Tom Coppens is Associate Professor of Urban Planning at the University of Antwerp and coordinator of the master programme in urban planning and design. He is an expert in planning instruments and spatial governance processes in Flanders and opinion maker on urban planning policy. His research focus lies on tradable development rights and planning instruments for the reallocation of building rights.

Chapter 10
Commentary: Swapping Development Rights to Prevent Flood Plain Development in Flanders: A Legal Architecture Perspective

John Sheehan

Introduction

Nature-based solutions for FRM are unavoidably impacted by the particularity of overarching national property laws and, especially, local property ownership and land use patterns. Crabbé and Coppens (this volume) in their paper "*Swapping development rights in swampy land: strategic instruments/strategies to prevent flood plain development in Flanders*" have chosen to respond to the question of particularity of property laws through the lens of flood plain development in the northern Belgian region of Flanders. In reflecting on the question of particularity in as much as it relates to NBS in FRM, I have also chosen a distinct approach that, however, eschews (without any overt or covert judgement) the empirical approach of Crabbé and Coppens.

This commentary is therefore imbedded with theoretical underpinnings of property rights that are, in any event, arguably and, indeed necessarily, informed by pragmatic legal considerations. To that end, the commentary changes gears but does not dismiss the road map evidenced in the empirical approach adopted in the Flanders case study. There are no broad rules or generalizations for NBS for FRM, and hence any conclusions must be absolutely neutral and objective. Crabbé and Coppens refer to the overall impact of the *gewestplannen* on land prices, and the resultant difficulties encountered in any changes to planned land uses presumably to address flood risks. Different disciplines necessitate consideration of FRM from a particular standpoint. Crabbé and Coppens appear to come from such a particular standpoint, providing an empirical consideration of FRM through an examination of the various opportunities available for governance of flood planning management through either

J. Sheehan (✉)
Faculty of Social and Economic Studies, Institute for Economic and Environmental Policy (IEEP), J. E. Purkyně University, Ústí nad Labem, Czech Republic
e-mail: John.Sheehan@uts.edu.au

Faculty of Society and Design, Bond University, Gold Coast, Australia

T. Hartmann et al. (eds.), *Nature-Based Flood Risk Management on Private Land*,
https://doi.org/10.1007/978-3-030-23842-1_10

market-driven instruments or, alternatively, government-led initiatives; essentially a land use planning discourse.

The issue of selection of the appropriate FRM governance models is, however, not exceptional, yet neither is the process of selection commonly encountered. Perhaps the task of ascertaining NBS for FRM necessitates an unexpected or even unanticipated focus on the process of selection of suitable and efficacious governance models in such specific circumstances. The impact upon private property rights ought not to be underestimated, and hence, from my perspective, the legal architecture surrounding any NBS for FRM is the touchstone for effectivity and efficiency. Whilst not suggesting any disrespect to other disciplines, the legal architecture could be described as the *liet motif* of any NBS for FRM on private land. Indeed, Crabbé and Coppens recognise this aspect to some degree when they note the reality of the oversupplied Flanders marketplace, which is adversely impacting the utility of possible transferable development rights for FRM. They also note mandatory land adjustment for FRM is similarly infused with the reality of the value of land in the hands of private owners who are unwilling to lose any value through the adjustment process.

Aside from FRM, even globally significant endeavours such as archaeology have also been impacted by the value of private land. For example, the government-sanctioned excavation in the 1890s by French archaeologists of the ancient Greek sanctuary of Delphi was delayed by some of three hundred dispossessed villagers who argued for better compensation for their private land beneath which lay the (now) UNESCO World Heritage Site (Cline 2017, 178). Unsurprisingly, with private property rights whirling around prospective NBS for FRM, the legal architecture is revealed as paramount.

Legal Architecture

Crabbé and Coppens determine that NBS for FRM in the face of perhaps significant financial compensation appropriately accruing to holders of private property is the major challenge for municipal, provincial and regional governments in Flanders. However, an understanding of the impact of NBS for FRM upon private property rights must first be nominally grounded in the constitutive conditions—social, legal, and institutional. Then secondly, the specific territorial composition of the conditions will subsequently confer legality and legitimacy upon those remedies considered appropriate to the particular territory. Such a starting point transposes the common narrative of the scientifically-informed foundation of FRM to a very different narrative of territorial politico-legal innovation within the reality of the driver of change, such as climate (Van Straalen et al. 2018, 190). The legal architecture of flood-prone areas in Flanders (and indeed, elsewhere) provides an important necessary impulse for the selection of the appropriate policy (instrument) to achieve the eagerly sought after NBS for FRM. Hence, the selection process is imbedded with the recognition that much flood-prone land is more often than not held as private property rights. I would argue that the common analytical distinctions between public environmental law and the law of private property ought not lead to presumptions that the legal architecture is so separate as to be irreconcilable, especially when tasked crucially with identifying a methodology (or methodologies) for FRM.

Indeed, whether the overarching legal architecture is based in a civil code regime such as in Belgium or a common law regime such as in England appears increasingly irrelevant, notably in environmental law. Convergence of intent allied with international law has increasingly overcome lingering conceptual historic obstacles thereby rendering omnibus environmental law recognizable and legitimate (Girard 2016; Needham et al. 2018, 53). Beyond Europe, this increasing irrelevance of legal genealogy is also manifested in the Asian Pacific region where notwithstanding the marked legal and political diversity of regional nation states, Couzens and Stephens detect "a number of common strands in the way in which governments in the region are approaching environmental challenges" (Couzens and Stephens 2017, 1).

Returning to the case study, the constitutive conditions—social, legal, institutional—clearly empower the municipal, provincial and regional governments in Flanders, and yet an overarching politico-legal remedy for NBS for FRM especially on private land appears to be remote. Firstly, why is this so? Secondly, why there are such almost intractable difficulties with implementation of broad-scale NBS for FRM in Flanders (and elsewhere)? From a property rights standpoint, the answer to both questions lies in a failure to appreciate any nature-based solution for FRM on private land is necessarily imbedded with theoretical underpinnings of property rights informed by pragmatic legal considerations. Perhaps, such a response merely reveals how the legal mind approaches the requisite accommodation flagged between FRM and private property rights. Possessive individualism lies at the heart of the theory of private property rights in respect of which Macpherson stated: "human society is essentially a series of market relations" (Macpherson 1962, 270). Nevertheless, Crabbé and Coppens appear uncomfortable with alternatives to government-led initiatives for FRM such as market-driven instruments notably transferable development rights or mandatory land readjustment. Yet, such alternatives represent a suite of tools that, in appropriate circumstances, can respond effectively to the presence of private property rights and address the significant financial worth of such privately held rights. Nonetheless, it is known that alternatives to government-led initiatives cannot serve as panacea for all management needs when flood risk has to be confronted.

Whilst somewhat prosaic, it is important to recognise NBS for FRM necessitate intervening in private property rights, and, ordinarily one can anticipate that to need significantly more land than more traditional flood risk "grey infrastructure" would require. Nevertheless, in a heartening observation Freyfogle stated: "private property, in fact, has been an evolving, organic institution with ownership rights that have varied greatly from era to era and place to place. The vast potential for further change of this institution very much needs exploring" (Freyfogle 2003, 7).

Research and Practice Questions Arising

Unsurprisingly, private property rights have always presented a conundrum for liberal-democratic states such as Belgium, Australia, or elsewhere. Freyfogle has further stated, "one has to do with the mismatch between the way private land is portrayed in law and culture and the way it exists in the real world of nature" (Freyfogle 2003, 7). This dual view of land, which includes private property rights, reveals national property law can only be ignored perilously by other disciplines when contemplating such issues as NBS to FRM on private land. The solutions sought after will only be realized through a cross-disciplinary collaboration between the perspectives offered by law and other disciplines where private land is to be the focus of the solution. The Flanders case study reveals existing attempts to introduce NBS for FRM utilizing private land frankly remain inefficient and probably in the main, ineffective. Because the question of particularity of property law impacts so severely on the prospect for efficiency and effectiveness, I consider the Flanders case study clearly reveals a need for urgent collaborative cross-disciplinary research into the theory and practice of flood risk mitigation and adaptation measures, notably where intervention in private property rights is required.

Acknowledgements Open access of this chapter is funded by COST Action No. CA16209 Natural flood retention on private land, LAND4FLOOD (www.land4flood.eu), supported by COST (European Cooperation in Science and Technology).

References

Cline EH (2017) Three stones make a wall: the history of archaeology. Princeton University Press, Princeton

Couzens E, Stephens T (2017) The prospects for a truly regional Asian Pacific environmental law? Asia Pac J Environ Law 20:1–4

Freyfogle ET (2003) The land we share: private property and the common good. Island Press, Washington DC

Girard F (2016) La propriete inclusive au service des biens environnementaux Repenser la propriete a partir du bundle of rights. Cah Droit, Sci & Technol 6:185–236

Macpherson CB (1962) The political theory of possessive individualism: Hobbes to Locke. Oxford University Press, Oxford

Needham B, Buitelaar E, Hartmann T (2018) Planning, law and economics, 2nd edn. Routledge, London

Van Straalen FM, Hartmann T, Sheehan J (2018) Conclusion: the social construction of changing environmental conditions. In: van Straalen FM, Hartmann T, Sheehan J (eds) Property rights and climate change: land use under changing environmental conditions. Routledge, London, pp 182–190

Prof. John Sheehan is currently Adjunct Professor in the Faculty of Design Architecture and Building at University of Technology Sydney. He is also Adjunct Professor in the Faculty of Society and Design at Bond University, Gold Coast, Queensland. In 2017, John was appointed Guest Professor with the Institute for Economic and Environmental Policy (IEEP) in the Faculty of Social and Economic Studies at J. E. Purkyne University in Usti nad Labem, Czech Republic. In 2003 he was appointed the independent Chair of the Project Advisory Committee Water Property Titles Program, Land and Water Australia, which was a research project funded by the Commonwealth Government to establish a water titling system for Australia.

Chapter 11
Commentary: An Economic View on the Prospects of a Flood Defense Related Development Rights Market in Flanders

András Kis and Gábor Ungvári

Economic instruments have been increasingly favoured by decision makers in the field of environmental and water policy, as they can provide efficient, low-cost solutions compared to command and control or traditional legal instruments. Such instruments, however, are not without problems. Even relatively simple solutions, for instance, environmental taxes, can backfire if their design, monitoring or enforcement is problematic, while more sophisticated instruments, such as market-based solutions, need even more thoughtful design and implementation to become successful. Some instruments look attractive on paper but are cumbersome or not operational in reality. The Flemish attempt at tradable development rights provides a suitable illustration of the practical difficulties of introducing an economic instrument that seeks to solve a problem through the creation and operation of a new environmental market.

The Origins of Demand and Supply

In Flanders, an economic solution to changing zoning designations was sought because government-initiated, forced rezoning had proven to be tedious and slow and require large financial compensations. In the proposed market, the development right can be decoupled from the land that is destined to be rezoned, and this right can be purchased by other landowners who are located in areas where development is not restricted. The traded commodity is the development right itself. The problem with this solution from an economic perspective is the gap between supply and demand in the case of Flanders. Supply can be offered by landowners who are located in flood-prone areas with restricted development options, whose development rights

A. Kis (✉) · G. Ungvári
Regional Centre for Energy Policy Research, Budapest, Hungary
e-mail: andras.kis@rekk.hu

G. Ungvári
e-mail: gabor.ungvari@rekk.hu

T. Hartmann et al. (eds.), *Nature-Based Flood Risk Management on Private Land*,
https://doi.org/10.1007/978-3-030-23842-1_11

were granted many years ago. In these areas, land with development rights is much more expensive than land without such rights, and this difference is the perceived value of the right itself that landowners would like to fetch in case they sell it.

Demand is generated by landowners outside of flood-prone areas who would like to develop their land. These actors, however, usually have other, cheaper alternatives than buying development rights from areas that are to be rezoned. As Crabbé and Coppens suggest, many developers can acquire such rights from their respective local authorities without much hassle and at a relatively low price. Therefore, there is little demand for development rights, and even that demand is priced low. No wonder that supply and demand cannot really meet.

How Could the Government Stimulate the Market?

In such a situation, an economist may wonder if demand and supply can be bridged through government intervention. A specific sort of intervention could be of administrative nature—restricting the issue of new development rights throughout Flanders for a set period of time, in order to provide incentives for the market-based transfer of these rights from flood-prone areas to other locations. The feasibility of this option is not clear to us, but we imagine it may create much dissatisfaction among developers.

Another alternative is that of a subsidy. Since the traditional approach is for the government to pay landowners to give up their rights, the government may as well pay market participants to facilitate transactions in the development rights market, thus reducing the price for the buyer or increasing the revenue for the seller. As the payment by the government can be lower than the usual practice of compensation based on land prices, the government can also benefit from this solution.

A related critical question, however, is not addressed in the paper: what is the value of the flood safety generated by rezoning the targeted land parcels? This value should act as a ceiling to any payment by the government, whether that payment is a direct compensation to landowners, or a subsidy to facilitate development right market transactions. There is discussion of the average expected annual damage due to flooding (50 million EUR/year in Flanders), but how much of this could be averted through land rezoning, is not clear. This information may not exist at all; however, it would be important to research it, as it would make an essential contribution to economically sound government flood policy.

An Alternative Solution: Auction-Based Purchase of Rights

Lastly, let us mention another alternative to the market of development rights, since its implementation obviously faces serious challenges. The government could consider auctions to purchase development rights from landowners in flood-prone areas. Participation by landowners would be voluntary, and participants would offer their development rights at a price that they consider high enough from their own perspective (their "reservation price"). Since different quantities would be offered at different prices, essentially a supply curve of development rights would be created, assuming that enough landowners participate in the auction. Demand by the government, on the other hand, would reflect the value of the public benefits of reduced flood risk, acting as a price ceiling at the auction. If the government makes it clear that these auctions represent the new main policy for rezoning, then landowners will start to offer their development rights since their revenue would be immediate and certain, while full future compensation based on real estate prices is uncertain. This type of certainty would likely help to drive down the purchase price of the rights. Regular auctions, such as annual ones, would be helpful: they would spread the financial burden of the government to several years while at the same time building trust in this new regulatory instrument and allowing landowners to learn about the auction mechanisms by studying already completed auctions. There are various auction types; the actual choice obviously would require a deeper analysis of the Flemish situation.

Some Prerequisites to Successful Environmental Markets

One of the lessons of this case study is that economic instruments are attractive in theory, but their practical implementation is not always viable, especially in the case of a legally complicated matter such as rezoning of inherited rights. There are also other obstacles, such as the standardisation of the traded product, connecting buyers and sellers, keeping transaction costs low. The number of trading transactions is also of importance: a relatively low number would probably not justify the efforts of establishing a market with all of its institutional requirements. In such a situation, looking for alternative solutions makes sense, including other, less complex economic instruments through which the economic efficiency could still be enhanced.

Another lesson is that developing market-based solutions requires a thorough prior understanding of the envisioned market by investigating questions such as the following: Who are the buyers and sellers? What is their motivation? What alternatives do they have compared to entering the market? Is the product to be traded well defined and understandable for all stakeholders? Would supply and demand likely meet—in other words, is there a scarcity that would drive supply and demand toward each other? Would there be enough participants in the market to make it liquid? How can transparency be ensured? Which institution would be in charge of registering transactions? How would administrative costs be financed?

Finally, even comprehensive research and preparation may not be sufficient to properly design economic instruments from the beginning. The experience with water markets (e.g., in Australia, USA), emission markets (most prominently carbon markets) and effluent discharge trading schemes (especially in the USA) promptly illustrates that a large part of the success of these solutions lies in their ability to evaluate participant feedback, adapt and evolve: learning by doing is often an inevitable component of realising the efficiency gains offered by economic instruments.

As we can see, the efforts required to create well-operating markets are substantial. The potential benefits, however, are also huge: properly designed and implemented market-based instruments can activate a large number of transactions, delivering valuable economic gains to a wide circle of buyers and sellers.

Acknowledgements Open access of this chapter is funded by COST Action No. CA16209 Natural flood retention on private land, LAND4FLOOD (www.land4flood.eu), supported by COST (European Cooperation in Science and Technology).

Andras Kis is a water economist at REKK, a think tank at the Corvinus University of Budapest, Hungary. He has 20 years of experience working as an environmental economist in Hungary, Central and Eastern European countries and the Middle East North Africa region. For the past 10 years his main focus has been water economics, including water utility regulation, assessment of economic policy instruments, cost-benefit analyses of flood protection measures, and benchmarking of water services.

Gábor Ungvári is senior water economist at the water economics unit of REKK in the Corvinus University of Budapest. Since 1999 he has worked in a series of water-resource focused, interdisciplinary research and economic consultancy programs. His analytical work (among many) targets FRM and the regulatory aspects of WFD measures. He is a member of the Scientific Council of the National Water Directorate of Hungary.

Part V
Wetlands Restoration in Czech Republic

Chapter 12
Urban Wetlands Restoration in Floodplains: A Case of the City of Pilsen, Czech Republic

Jan Macháč and Jiří Louda

The chapter deals with the importance of implementing small-scale NBS for flood protection in cities. It turns out that, although NBS bring multiple environmental and social co-benefits, their real-world implementation comes up against numerous barriers, particularly if private land is needed for their implementation. Insufficient awareness of the benefits of such measures reduces stakeholders' willingness to consider their implementation (including negotiations with private owners about provision of land for these purposes). If it is necessary to use private land for NBS implementation in cities (thus public funds for compensations to the private owners), the social benefits of such measures have to be demonstrated to stakeholders in a clear and transparent manner. In regard to the case study of Lobezská louka, where an NBS in the form of urban wetlands has already been partly implemented, we demonstrate the potential of economic assessment in the form of a cost-benefit analysis to facilitate decision-making for further application of NBS in cities. In our case, the analysis showed that the social benefits of the measure realized in the first phase exceed the costs 25 times. The result is a supporting argument for implementation of the next two phases on private land.

Introduction

Urban areas are among territories where manifestations of climate change have very significant adverse (economic) impacts, and adaptation to such change (in the form

---

J. Macháč (✉) · J. Louda
Faculty of Social and Economic Studies, Institute for Economic and Environmental Policy (IEEP), J. E. Purkyne University in Usti nad Labem (UJEP), Ústí nad Labem, Czech Republic
e-mail: machac@ieep.cz

J. Louda
e-mail: louda@ieep.cz

T. Hartmann et al. (eds.), *Nature-Based Flood Risk Management on Private Land*,
https://doi.org/10.1007/978-3-030-23842-1_12

of more frequent floods or droughts, heat waves or urban heat islands) incurs considerable social costs.

Until recently, "traditional" adaptation of cities to climate change has relied mostly on technical measures ("grey infrastructure") such as mobile flood barriers, canalisation/culverting of watercourses, construction of dams and dikes (upstream of the city), insulation of building envelopes, and installation of air-conditioning. Alternative solutions to technical measures are NBS, which can be seen as complements to technical measures and can be implemented directly in the urban area. "The fundamental belief is that NBS can represent more efficient and cost-effective solutions than traditional approaches" (Lafortezza et al. 2017). One of the reasons is that implementation of NBS may bring multiple co-benefits (McVittie et al. 2018) improving the well-being of inhabitants. More widespread application of NBS in cities also has strong support from the European Commission (Faivre et al. 2017). Ever more cities thus reach for NBS, although their implementation does face numerous barriers. Once of them is the space requirement of NBS. Compared to grey infrastructure, these measures tend to interfere with privately-owned land more often.

This chapter is focused on the main challenges of and obstacles to NBS[1] implementation as part of flood event adaptation in cities from an economic and institutional point of view (assuming that the city's goal is to build a conceptual system of flood protection). From this point of view, the main question can be defined as follows: How to support the planning and implementation process of NBS in cities, especially if they should be implemented (at least partly) on private land? One possible solution is demonstrated on a case study of the city of Pilsen. This city decided to implement NBS to solve its flood protection.

In general, awareness about NBS and the benefits brought by their implementation is limited. Solving this issue involves three different levels of stakeholders:

- decision makers (especially local politicians),
- private landowners,
- public administration, NGOs and other stakeholders.

Implementation of flood protection measures (and especially "novel" measures such as NBS) depends heavily on support from decision makers and other stakeholders. This means that one of the main challenges is to convince the municipal government of the benefits and importance of NBS implementation, because members of the municipal government make the decisions on spending the public budget. Politicians prefer short-term results compared to long-term ones due to the political cycle and the effort to be re-elected. In the current term, the decision makers have shown preference for measures related to recreation. Moreover, in the last three years, the Czech Republic has been more affected by droughts than by floods. Floods are thus not perceived as a significant risk, or at least not by the current municipal representation. This is in contradiction with the extensive damage caused by floods in the previous 20 years.

[1]Or sometimes referred to as nature-based flood protection measures (NFPM).

Another important issue for a city's successful adaptation is related to private ownership of land in cities. A large part of land in urban areas is owned by private entities. Designing and implementing measures on (previously) private land seems to be a key challenge on the road to a successful adaptation process. Decision makers are restrained from negotiating with private owners, because transaction costs and compensations to landowners for using the land for NBS increase the time and costs of implementation.

A third group of stakeholders is composed of entities that are not directly involved in the process of NBS implementation but may significantly influence the results of negotiations about new measures, such as creating a new urban wetland. NGOs often have a strong position in the civil society and can influence local inhabitants' perception regarding NBS. On the other hand, many entities in public administration (such as a municipal environmental or urban development department) are either directly involved in the decision-making process, or prepare inputs for the municipal government.

The rest of the chapter is organised as follows: the second section presents the broader context of flood-related issues in the city of Pilsen followed by a detailed description of the case study (Lobezská louka urban wetlands restoration). The third section very briefly presents the three perspectives that have to be taken into account in the NBS planning and implementation process (hydrology, economics and stakeholders). The following section is focused on one of these perspectives: economics of NBS. The last section discusses the role of economic analysis (comparing costs and multiple co-benefits of NBS) in the process of NBS implementation in cities, especially where private land is needed for measures.

Pilsen Case Study: Small-Scale Nature-Based Solution—An Answer to Regular Flooding of the City

The city of Pilsen (Plzeň) is situated in the western part of the Czech Republic, with a population of 170 000, making it the fourth biggest city in the country. Four smaller rivers flow through the city (Úhlava, Úslava, Radbuza and Mže) and form together the Berounka river. Water has historically been one of the main factors influencing the development of the city of Pilsen. On the one hand, the water was used for centuries as a fundamental source for industrial development (brewery, heavy industry, etc.); on the other hand, water has brought regular destruction to the city.

Floods are one of the most serious threats to the city in relation to climate change. The flood zones of the four smaller rivers are depicted in Fig. 12.1. Pilsen has dealt with floods on a regular basis—there have been seven big floods between 2002 and 2014. In 2002, Pilsen was one of the cities in the Czech Republic most affected by floods: 11% of the city's area was flooded, and damages exceeded EUR 21.5 million.

Based on this experience, Pilsen has faced the challenge of finding a solution to flood damage. The public demand for flood risk reduction was high, but was mostly

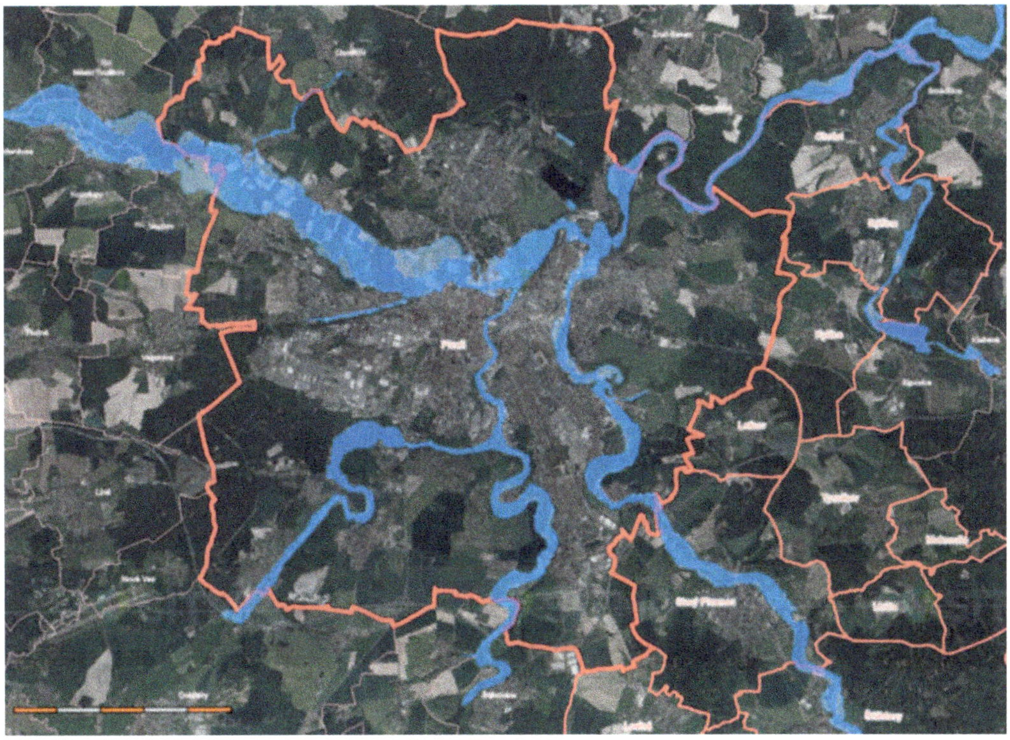

Fig. 12.1 Flood zones in the city of Pilsen (SITMP 2017)

concerned with large-scale structural measures (building dams or dikes) or small-scale technical measures such as demountable flood protection barriers. The same was true for policymakers, who were concerned about the issue, but not informed about all possible solutions. Implementation of large-scale flood protection measures (dikes and other measures based on grey infrastructure) mostly extends beyond the city's cadastral area and requires accord of a wide range of stakeholders at the trans-municipal and trans-sectoral level. This kind of comprehensive planning should take into account the upstream-downstream relations (Macháč et al. 2018), and it is a long-term process. The confluence of four rivers complicates the problem because regular flooding cannot be resolved by one large-scale upstream measure, but instead it would be necessary to implement measures on each tributary. Thus, decision makers in the city of Pilsen decided to implement small-scale NBS inside the city.

The process of designing and implementing a system of NBS for flood protection (consisting mainly of urban wetlands) started in Pilsen in 2008, when most of the measures were designed within the international project REURIS (2008–2011). Later on, many of the NBS designed (e.g., revitalisation of Božkov Island, creation of the park Lobezská louka with four wetlands and Pod Vyšehradem wetlands) were implemented continuously in the period 2009–2015 (partly financed by EU funds—Operational Programme Environment). By implementing these measures, the city of Pilsen pursued four main objectives: (i) design of appropriate nature-based flood protection measures, (ii) revitalisation of the river areas (streams and floodplains), (iii) finding suitable utilisation for a large area of floodplains and river banks, (iv) creation of conditions for implementing the territorial system of ecological stability.

At the present time, the municipal system of urban wetlands in Pilsen extends over an area of 14 ha. In this first phase, the measures have been mostly implemented on land owned by the city, without the need for negotiation with private owners. A large number of other measures that have not yet been implemented are located on privately owned land. For the next phases, the challenge of privately owned land that is suitable for further NBS should be solved.

Although the system of NBS in Pilsen is a small-sized green infrastructure solution with relatively little impact (from the flood protection point of view), it is combined with a recreational function, and hence it brings multiple benefits for the community (see later on). This kind of information could play a crucial role in the decision-making process on implementing of further measures, especially in case private land is needed for their implementation.

Small-Scale Nature Based Solution Implementation in "Lobezská Louka" Park

In the case study, we would like to present a method of expressing the societal benefits of urban NBS and use this information for supporting further development of similar measures also on (previously) private land.

The "Lobezská louka" park with four wetlands is one of the NBS implemented in the city of Pilsen. It was formerly an area of neglected greenery; the original unmanaged green areas on the site gradually overgrew with herbs and tree seedlings, and illegal dumps also occurred. It is located in the immediate vicinity of the Úslava river. Almost 10 000 people live in the neighbourhood of the area. The current "Lobezská louka" park is only a part of the measures designed for this area. The implementation of this first phase is the result of long-term efforts by the city of Pilsen for a suitable and modern design of the embankment of the Úslava river inside the city. According to Atelier Fontes (2010), the initial plan of possible measures covered an area of approximately 14 ha. Only a small part of the land is owned by the city, the rest by private owners. The owners have limited opportunities for land use due to the floodplain, the Q_{100} active zone and the land-use plan. With regard to ownership, the original project was divided into several phases (see Fig. 12.2). The project objective is to establish water bodies in the wide alluvial plain in order to both reduce the flood flow rates by means of a nature-based measure and to increase the aesthetic value of the area and make a place for recreation.

The total area consists of 44 plots of which 21 are in private ownership and 23 are owned by the municipality of Pilsen. The area of former horticulture (central part of the area) was selected as an appropriate site for the first phase. One of the reasons is that all the seven plots were owned by the municipality. The initial situation regarding ownership is captured in Table 12.1. Only 46% of the total area was owned by the municipality. The measures in the second and third phases have been planned as different options with regard to land availability. The final form of the NBS depends on negotiations with other private owners. The result may be different from the plan in Fig. 12.2.

The first phase covers an area of 2.0 ha. Initially, the first phase was designed on an area of 3.5 ha. The decision-makers decided to build wetlands only on land owned by the city. This step was chosen primarily for reasons of easier grant application. According to the requirements of EU funds—Operational Programme Environment—the owner has to agree with the implementation of measures. If the phase were to be built on private land, it would be necessary to buy out the land. The project implemented as the first phase involved the establishment of four wetlands with water retention potential (wetland biotope) and an adjoining park. In contrast with other "common parks" with a lake, the park in our case study differs in its range of services and benefits that it provides. The area is used daily by the local inhabitants. A specific feature of this area is that it combines functions of flood retention (wetlands), education (educational trail with information boards) and recreational functions, such as opportunities for swimming in these wetlands and other types of recreation (relaxation, sports, walking). A cycle path along the river that connects different parts of the city is also a part of the project. The wetlands have no feeder or drain canals and are only filled with groundwater and rain. The four wetlands (Fig. 12.3) have the capacity to accommodate around 8000 cubic metres of water. The first phase was completed in the summer of 2015. Implementation of the first

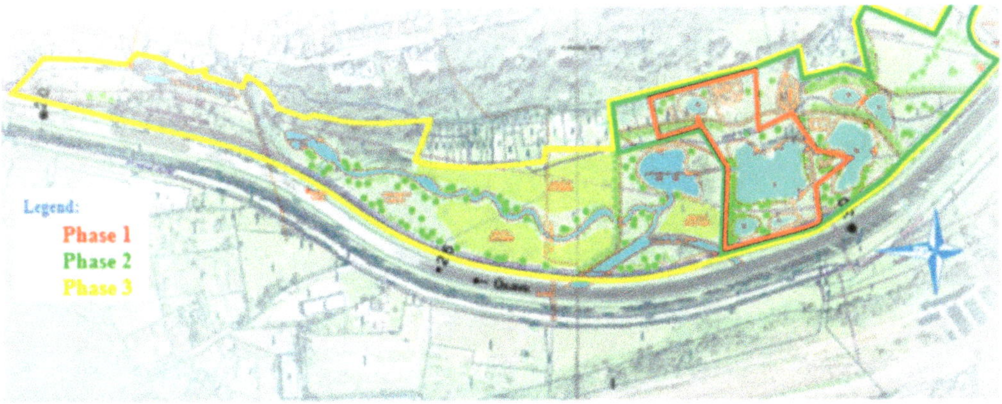

Fig. 12.2 Plan of the "Lobezská louka" area and division of measures into phases 1–3 (Útvar koncepce a rozvoje města Plzeň 2015)

Table 12.1 Land ownership regarding the phases

Owned by	Phase 1 (ha)	Phase 2 + Phase 3 (ha)	Total area (ha)
Municipality	2.0	4.5	6.5
Private owners	–	7.5	7.5

Source Based on Atelier Fontes (2011)

phase was not without problems. The groundwater was deeper than expected, which led to problems with filling the wetlands with water. This was caused mainly by a drought that affected the Czech Republic in 2015.

Currently, the revitalisation of a nearby forest park is under preparation. The implementation of the other two phases "Lobezská louka" is hampered by complicated property relations. It is necessary to buy out the privately owned land or to negotiate with private owners about utilisation of their plots. Currently, a part of the plots for the second and third phase has been bought out from private owners. At the moment, not enough support for the project has been obtained from the city council as only the local district town hall supports the realisation, but the city district does not have sufficient financial resources. The 4.5 ha of land owned by the city can be used for developing the NBS. Furthermore, it is essential to convince the public and decision-makers about the need for small NBS.

Three Perspectives of NBS Implementation

To answer the main question concerning support to the NBS planning and implementing process in cities (on private land), it is necessary to deal with three different perspectives: hydrology, economics and stakeholders (see Fig. 12.4). Each of them plays an important role in the process of NBS implementation.

Fig. 12.3 Urban wetlands "Lobezská louka" in Pilsen (Provided by Útvar koncepce a rozvoje města Plzeň)

The first perspective that has to be assumed in the process of NBS implementation in cities is a hydrological analysis, which produces a description of the current hydrological conditions and helps to design appropriate and efficient measures in the context of physical flood protection.

From the point of view of economic analysis, we distinguish between two main aspects: the financial aspect and the aspect of (net) social benefits resulting from implementation of the measures. The financial aspects are closely interconnected with those individuals who decide on the available budget for implementing measures and are able to ask for additional money from the state budget in the form of subsidies. The other aspect of economic analysis is focused on assessment of the net social benefits resulting from the measures. Net social benefits can be used for argumentation in the social debate about implementation of NBS, because besides a primary purpose (flood protection in our case), applications of NBS provide further multiple environmental benefits (Nesshöver et al. 2017) such as improvement of air quality, habitat for species, soil erosion control, etc., and other economic and social benefits (e.g., increase in aesthetic value, energy/water cost savings, etc.).

The last perspective includes all stakeholders and especially decision makers and landowners, whose decision holds the most weight about the implementation. Their awareness of NBS benefits plays the most important role in the implementation process. A stakeholder analysis is an appropriate tool to identify their preferences and attitudes but also barriers and challenges towards water management measures (especially NBS), taking into account different points of view of particular stakeholder groups.

The second and last perspectives are important for the problem of privately owned land. There are significant synergies between these two perspectives (economic argument and communication between stakeholders), which can motivate both stakeholders and decision makers to agree with NBS and the implementation.

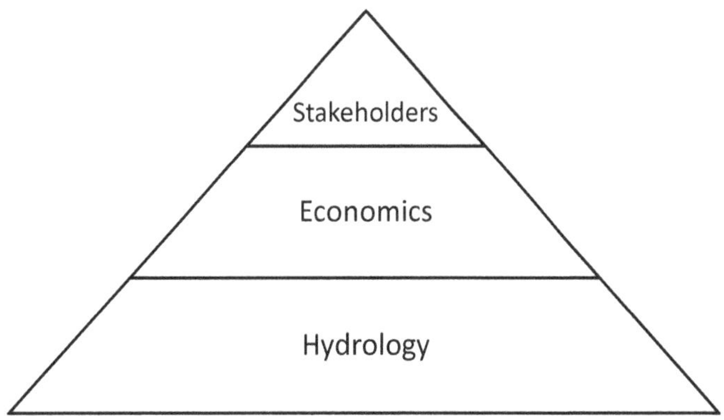

Fig. 12.4 Perspectives of NBS implementation in cities

The Need for Strong Arguments

In our case, the stakeholder analysis showed that perception of environmental issues differs across the stakeholders (likewise, the perception of different NBS), but there was agreement on the point that flooding is the most important problem in Pilsen. One barrier to tackling floods in the city more effectively is that less area is available than required for effectively implementing measures (availability of vacant municipal land, complicated property relations outside municipal land, high prices of private land). Another crucial barrier to NBS implementation (on private land) is a lack of political support (thus, lack of funds for implementation of such measures), although many officials from related municipal departments are interested in innovative solutions (such as construction of wetlands in cities). The low level of awareness regarding the importance of NBS and their benefits leads to their marginalisation as a meaningful solution to the issues.

According to McVittie et al. (2018), “the demonstration of multiple co-benefits may be important both where these co-benefits provide private benefits for land managers and wider societal benefits that can attract a variety of funding sources” and political support. We assert that economic analysis (cost-benefit analysis in our case) and appropriate communication of its outputs in the NBS planning phase could help to overcome the above-mentioned barriers, particularly the problem of insufficient awareness among politicians and other stakeholders about the (direct and indirect) benefits of these measures.

As mentioned above, a complex (but rather small-scale) system of urban wetlands was designed on the site of our case study, and its implementation was divided into three phases. Thanks to financial support from EU funds and because the land needed for the first phase of implementation was in municipal ownership, the first four wetlands together with revitalisation of part of the park were implemented. However, the land for the remaining two phases is mostly in private ownership. Since new urban wetlands do not bring net benefits for private owners, implementation of these measures will require additional financial resources from public budgets. In addition to the investment costs for implementing measures, the landowners should be compensated for providing the land (e.g., buyouts, long-term leases or some kind of so-called payments for ecosystem services). To persuade the decision makers (mostly municipal politicians) and other stakeholders to implement the further phases of urban wetland construction in the “Lobezská louka” area, there is a need for strong arguments about the multiple benefits provided by NBS for the whole society.

To demonstrate the environmental and social benefits of urban wetlands restoration, the first phase of the “Lobezská louka” project was valuated. The assessment of the society-wide benefits of the measure implementation was based on the economic cost-benefit analysis (CBA) method, which takes into account not only private financial benefits and costs of the implementing entity, but also the costs and benefits resulting for society as a whole (non-financial and indirect costs and benefits). In addition to the primary benefit consisting in direct contribution to flood protection (flood risk reduction), NBS bring numerous co-benefits contributing to the populations' well–being (e.g., property value increase, support to biological diversity, spaces for recreation and meditation, etc.). The identification of benefits is based on the ecosystem services approach. Besides ecosystem services divided into 4 groups (supporting, regulating, provisioning and cultural services; see Fig. 12.5), other benefits such as biodiversity (habitat creation) were also taken into account.

A cost-benefit analysis consists of several steps (see Fig. 12.6). In the first step, the evaluated measure is described. In the next step, individual costs and benefits are identified using the concept of ecosystem services. Benefits are quantified using biophysical indicators and expressed in monetary value using appropriate methods. The costs are set according to project budgets (investments costs) and estimated operating costs.

The comparison of costs and benefits used the annualised value of costs and benefits. The concept of annualised costs and benefits is derived from the concept of real value of money and the opportunity to invest funds elsewhere (Jacobsen 2005). The known present costs and benefits are transformed into a future flow of the same values based on annual costs, which (when cumulated) match the known present value.

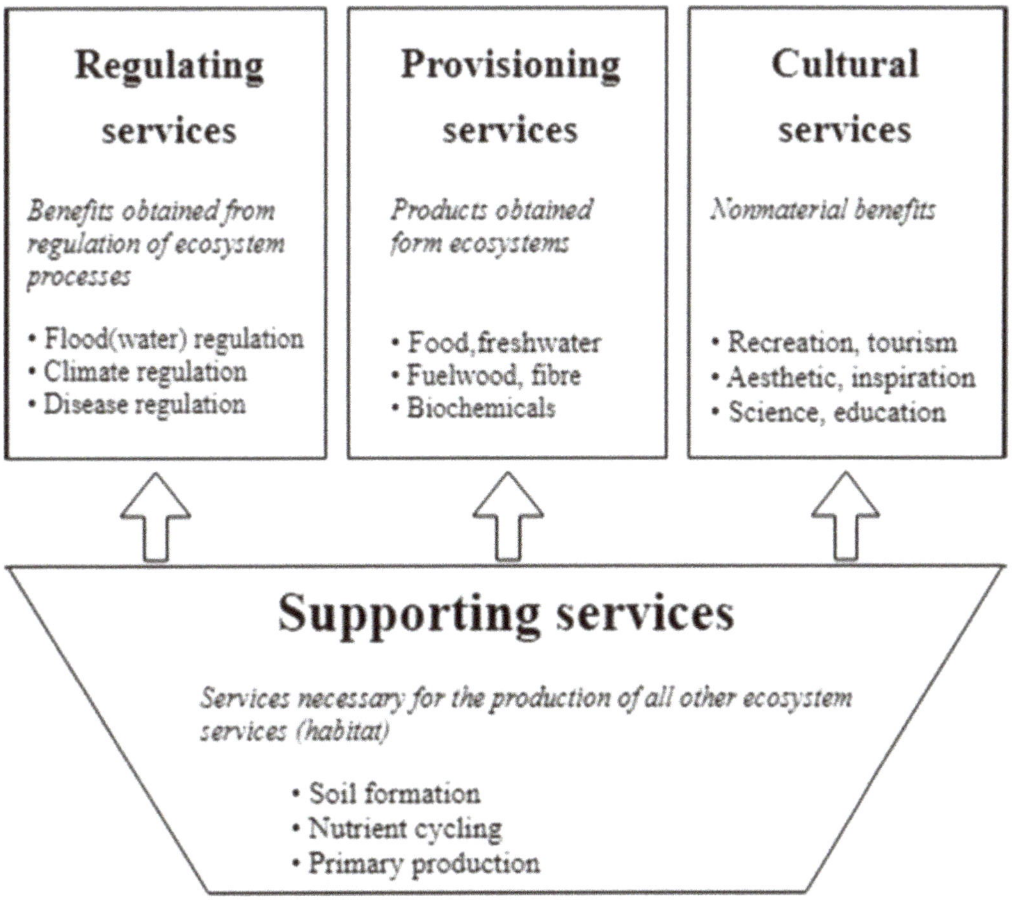

Fig. 12.5 Ecosystem services connected with NBS (based on Millennium Ecosystem Assessment 2005)

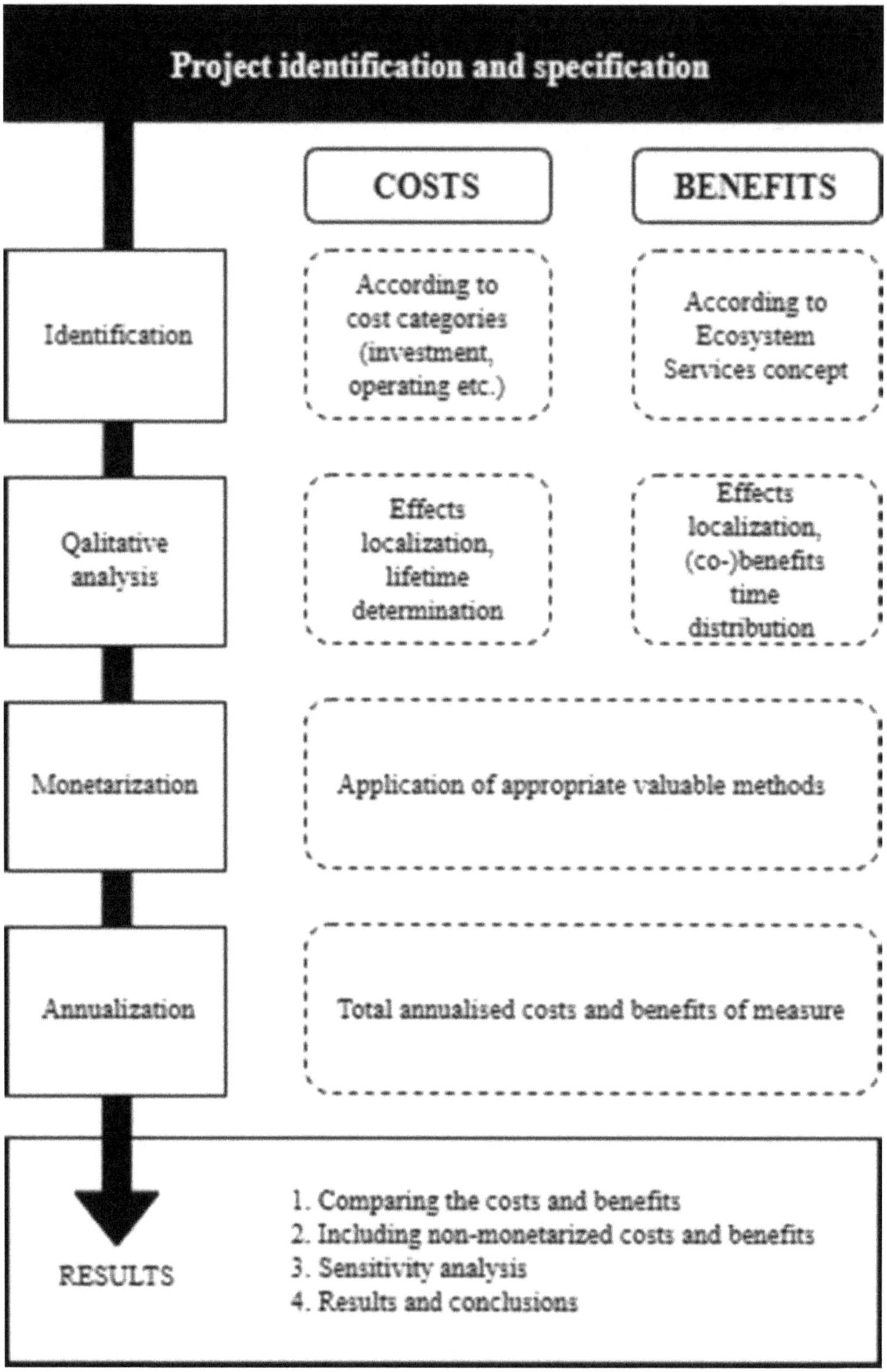

Fig. 12.6 Cost-benefit analysis step by step (based on Slavíková et al. 2015)

Due to the wide range of ecosystem services provided by NBS and the lack of primary data, the benefit assessment used the benefit transfer method in the form of a meta-analysis, which makes it possible to use secondary data from similar sites and transfer them to the area being assessed while involving local conditions in the assessment (Brouwer et al. 1999).

In addition to flood protection benefits, applications of urban wetlands provide a number of co-benefits in terms of ecosystem services. The assessment led to identification of 14 major services/benefits, but not all of them were valuated in monetary terms (see Table 12.2). Regulating services are the most numerous. Benefit transfer was used for valuation of 8 services to quantify the annual benefit of both the wetlands themselves (EFTEC 2010) and the adjacent park greenery (Patrick and Randall 2013). The data transfer takes into account the primary analysis methods, GDP in the area, population count, distance from the centre, number of similar wetlands in the surrounding area, etc.

After consideration of local aspects, the ex-post CBA of the first phase of the urban wetland restoration showed that the annual benefits of this measure amount to EUR 1.47 million. The adjacent greenery contributes significantly to the total amount of benefits. In addition to direct impacts on water retention and reduction to flood damage, it has a noticeable influence on the recreational function and water and air quality. Only part of the services provided was quantified in monetary units by the benefit transfer application.

The cost valuation was based primarily on the investment costs of implementation of the wetlands themselves, as well as the operating costs of periodic maintenance and other irregular costs of management of the area. The total annual costs are about EUR 0.06 million. The annual costs include investment costs (EUR 0.526 million according to City of Pilsen data), operating costs of park greenery maintenance (EUR

Table 12.2 Identified ecosystem services provided by "Lobezská louka" wetlands

Type of benefit	Monetary valuation	Type of benefit	Monetary valuation
Reduced risk of flooding	Yes	Erosion reduction	Yes
Supply of surface water and groundwater	Yes	Real estate value	No
Improved water quality	Yes	Recreational benefits	Yes
Regulation of micro-climate/city heat island	No	Increase in aesthetic value	Yes
Noise reduction	No	Biomass production	No
CO_2 reduction	No	Crop production (urban agriculture)	No
Air quality improvement	Yes	Habitat creation	Yes

$0.41/m^2$/year) and other costs connected with the wetlands (periodic maintenance and the less periodic costs of desilting the wetlands).

When comparing the total annual benefits and costs, we can see that the benefits exceed the costs nearly 25 times. The annualised costs are only 4% of the total annual benefits. The total annual benefits have to be perceived as a monetary expression of the ecosystem services mentioned above. Thus, they are not benefits that could be followed in the form of cash flows for citizens or municipality. On the other hand, not all the environmental and social benefits were included in the monetary valuation; therefore, the final figure of benefits has to be regarded as underestimated. The significant excess of the social benefits over the costs is confirmed by the results of the sensitivity analysis, which tested the effect of the most important factors on the study results. The effect of the discount rate was tested above all.

Searching for a Way to Support Implementation of Nature-Based Solution in Cities

Solving the flood problem at the city level requires a comprehensive approach. Although large-scale technical measures such as dams often theoretically seem to be appropriate for preventing flood damage, it is seldom within the city's powers to implement such measures (dams, dikes and polders have to be built outside the city's cadastral area upstream the river, negotiations with other municipalities, catchment area administrator, etc., are necessary, and moreover they are very costly measures that cities refuse to fund outside their territories). Besides such large-scale measures, cities have the additional opportunity to focus on more local measures, which (compared to large-scale measures such as dams) are relatively fast to implement directly inside the city. In this respect, Czech cities often resort to the application of single-purpose "grey" measures, such as mobile flood dams. There is also a third option—small-scale nature-based flood protection measures, which can be (from the technical point of view) relatively easily implemented directly in city centres. Although the NBS bring multiple social and environmental co-benefits, their real-world implementation (at least in Pilsen, but we assume other cities as well) often comes up against numerous obstacles that are frequently related to lacking information about their society-wide benefits and to private ownership of land needed for NBS implementation. The stakeholders (and mainly municipal politicians) are very cautious about spending public money on measures with doubtful benefits (from their point of view).

In order to boost the awareness of the importance of such measures, it is not necessary to carry out a detailed economic analysis for each planned measure. The importance of measures can also be documented with successful examples from other cities or countries. The costs and the benefits expressed in monetary terms can be compared easily without having to understand the numerous direct and indirect benefits, which may lead to better awareness about the NBS and, subsequently, significantly help to implement them. This factor is also of considerable significance when implementing measures on land not owned by the municipality. The case study presented above can be used for that purpose.

In the "Lobezská louka" area in Pilsen, apart from previously implemented measures (Phase 1) paid largely with EU funds, many additional measures of this urban wetland system (Phase 2 and 3) are ready to be implemented (from the urban planning point of view). The land needed for their implementation is mostly owned by private entities. The knowledge of all the benefits that the measure will bring for the society may increase the city's willingness to negotiate on buying out the land and then use it for NBS implementation. Alternatively, application of some form of payment for ecosystem services may be considered (Kumar et al. 2014; Cerra 2017; Reed et al. 2017). In that case, the city would not buy out the land but only pay its owner a certain fee for providing the water retention service (such as in the form of lease or subsidy).

The outputs of ex-post CBA for the first phase of the "Lobezská louka" project proved that the social and environmental benefits of this measure are at least 25 times higher than the costs. This information may be used as a crucial argument for the social debate and decision-making process about implementing further phases of the urban wetlands in Pilsen (and especially in thinking about buyouts or other type of payments for the private land).

In general, we conclude that even though small-scale NBS in urban areas are not able to solve the whole problem of regular flooding, they can complement large-scale measures. NBS can bring significant environmental and social co-benefits compared to small-scale single-purpose technical measures. Social acceptability of the measures and their implementation depends on public awareness. That is why we argue that the co-benefits (environmental and socioeconomic) should also be considered when deciding. We assert that a key factor for supporting NBS implementation at the city level is improving, in a simple and transparent manner, information among stakeholders, particularly decision makers, about all the benefits that this type of measures provides for the society (and, if possible, about their monetary value).

The CBA results can be utilised not only in argumentation in favor of implementing the two remaining phases of the "Lobezská louka" project, where complex property relations will have to be resolved, but also in the case of implementing other nature-based measures in general. Application of the CBA method leads to the aggregation of all the benefits in a single figure. On the one hand, this blurs the importance of the different ecosystem services; on the other hand, it simplifies communication of the net benefits to the general public as it expresses everything with a single figure. In other words, it enables comparison of the financial costs of implementation and maintenance of measures with the benefits, which mostly lack direct financial impact but make a significant contribution to quality of life.

Acknowledgements We thank the Operational Programme Research, Development and Education of the Czech Republic for financing the project Smart City—Smart Region—Smart Community (grant number: CZ.02.1.01/0.0/0.0/17_048/0007435) that led to the present book chapter.

Open access of this chapter is funded by COST Action No. CA16209 Natural flood retention on private land, LAND4FLOOD (www.land4flood.eu), supported by COST (European Cooperation in Science and Technology).

References

Atelier Fontes (2010) Study Povodňový park. Basis for feasibility study. Atelier Fontes, Brno 19 pp

Atelier Fontes (2011) Plzeň – Lobzy: Povodňový park. Feasibility study. Appendix No. 9. Atelier Fontes, Brno

Brouwer R et al (1999) A meta-analysis of wetland contingent valuation studies. Reg Environ Change 1(1):47–57. https://doi.org/10.1007/s101130050007

Cerra J (2017) Emerging strategies for voluntary urban ecological stewardship on private property. Landscape Urban Plann 15:586–597. https://doi.org/10.1016/j.landurbplan.2016.06.016

EFTEC (2010) Valuing environmental impacts: practical guidelines for the use of value transfer in policy and project appraisal case study 3—valuing environmental benefits of a flood risk management scheme. report for Defra. London. DIALOG. https://www.gov.uk/government/uploads/system/uploads/attachment_data/file/182376/vt-guidelines.pdf

Faivre N, Fritz M, Freitas T, de Boissezon B, Vandewoestijne S (2017) Nature-based solutions in the EU: innovating with nature to address social, economic and environmental challenges. Environ Res 159:509–518. https://doi.org/10.1016/j.envres.2017.08.032

Jacobsen M (2005) Project costing and financing. In: Lønholdt J (ed) Water and wastewater management in the tropics. IWA Publishing, London, pp 51–119

Kumar P, Kumar M, Garrett L (2014) Behavioural foundation of response policies for ecosystem management: what can we learn from Payments for Ecosystem Services (PES). Ecosyst Serv 10:128–136. https://doi.org/10.1016/j.ecoser.2014.10.005

Lafortezza R, Chen J, van den Bosch CK, Randrup TR (2017) Nature-based solutions for resilient landscapes and cities. Environ Res. https://doi.org/10.1016/j.envres.2017.11.038 (in press)

Macháč J, Hartmann T, Jílková J (2018) Negotiating land for flood risk management—upstream-downstream in the light of economic game theory. J Flood Risk Manag 11(1):66–75. https://doi.org/10.1111/jfr3.12317

McVittie A, Cole L, Wreford A, Sgobbi A, Yordi B (2018) Ecosystem-based solutions for disaster risk reduction: lessons from European applications of ecosystem-based adaptation measures. Int J Disaster Risk Reduct. https://doi.org/10.1016/j.ijdrr.2017.12.014 (in press)

Millenium Ecosystem Assessment (MEA) (2005) Ecosystems and human well-being: Synthesis. Island Press, Washington D.C

Nesshöver C, Assmuth T, Irvine KN, Rush GM, Waylen KA, Delbaere B, Haase D, Jones-Walters L, Keune H, Kovacs E, Krauze K, Külvik M, Rey F, van Dijk J, Vistad OI, Wilkinson ME, Wittmer H (2017) The science, policy and practice of nature-based solutions: An interdisciplinary perspective. Sci Total Environ 579:1215–1227. https://doi.org/10.1016/j.scitotenv.2016.11.106

Patrick E, Randall A (2013) International meta-analysis of green space for benefit transfer. Contribution on conference: The Australian Agricultural and Resource Economics Society Inc. 57th Annual Conference, Sydney, Australia. DIALOG. http://www.aares.org.au/aares/documents/2013AC/Handbook.pdf. Accessed 5–8 Feb 2013

Reed MS et al (2017) A place-based approach to payments for ecosystem services. Glob Environ Change 43:92–106. https://doi.org/10.1016/j.gloenvcha.2016.12.009

SITMP (2017) Flood zones in the city of Pilsen. DIALOG. https://gis.plzen.eu/zivotniprostredi

Slavíková L, Vojáček O, Macháč J, Hekrle M, Ansorge L (2015) Metodika k aplikaci výjimek z důvodu nákladové nepřiměřenosti opatření k dosahování dobrého stavu vodních útvarů. Výzkumný ústav vodohospodářský T. G. Masaryka, v.v.i., Praha. ISBN 978-80-87402-42-9
Útvar koncepce a rozvoje města Plzeň (2015) Plan from information point about Lobezská louka wetlands

Jan Macháč is an environmental economist at the Institute for Economic and Environmental Policy (Jan Evangelista Purkyně University in Ústí nad Labem). He mainly focuses on implementation of NBS and the adaptation of cities to climate change and water management from an economic perspective. He is the author of a Czech certified methodology for economic assessment of green and blue infrastructure in cities. He graduated with a Ph.D. from the University of Economics in Prague.

Dr. Jiří Louda graduated from the University of Economics in Prague and is currently a senior researcher at the Institute for Economic and Environmental Policy (Jan Evangelista Purkyně University in Ústí nad Labem). His scientific activities focus on the application of ecosystem services concept in practical environmental and planning policies, especially at the municipal and regional level. Payments for ecosystem services in relation to NBS implementation is one of the key points of his research.

Chapter 13
Commentary: Urban Wetlands Restoration as NBS for Flood Risk Mitigation: From Positive Case to Legitimate Practice, in the View of Evidence-Based Flood Risk Policy Making

Jenia Gutman

From a policy development and implementation perspective in the light of adaptive governance, the case of wetland restoration in Pilsen raises issues of extreme complexity faced by positive case "enablers" when about to upscale from small-scale solution to acknowledged wide-scale outcome. Topics such as a higher degree of compliance with the core conventions of the EU Flood Directive (hydrologically sound, evidence-based planning and execution) as well as with other policy stepping stones are suggested and discussed.

Zooming in—From Flood risk Directive to NBS: Since the beginning of the 21st century, the persistent attempt to attack the "wicked problem" (Rittel and Webber 1973) of floods in a systematic manner has given rise to a variety of ideas and concepts. Integrated watershed (or catchment) management (IWM/ICM) is the keystone concept underlying the EU Flood Directive (2007/60/EC). In light of IWM principles, the Water Framework Directive (2000/60/EC) and the subsequent Floods Directive (2007/60/EC) refer to the watershed (catchment, basement) unit as the most appropriate framework to manage the ecological state of rivers and other water bodies as well as to manage flood risks. The reason behind this decision lies in the fact that the watershed is "just about right" for the intended function as it incorporates the entire water cycle in nature, the upstream and the downstream communities, the stressors causing or expanding the flood risk (FR) and the receptors of the flood risk (van Ruiten and Hartmann 2016). Only the watershed level allows full understanding of the structure and functioning of the eco-hydrological "big idea" on which the human activities are based. This "big idea" and the human dimensions in it allow for appropriate and resilient flood risk governance arrangements by informing where, why

J. Gutman (✉)
Department of Soil Conservation and Drainage, Ministry of Agriculture and Rural Development, Rishon-Lezion, Israel
e-mail: jeniag@moag.gov.il

T. Hartmann et al. (eds.), *Nature-Based Flood Risk Management on Private Land*,
https://doi.org/10.1007/978-3-030-23842-1_13

and what measures should be taken to enhance the resilience of social-ecological systems to flood risks (Dieperink et al. 2016). Though this holistic, watershed-based approach is not new, latest reports still highlight the need in bridging for improving upon fragmented flood risk measures (FRM) in the light of the entire watershed (Hegger et al. 2016a). One of the plausible reasons for absence of the watershed in the NBS discourse possibly lies in the fact that human senses cannot perceive watersheds, while NBS in urban areas are easily detected and experienced. A human watershed (Brunckhorst and Reeve 2006) is only a few blocks in size, corresponding to the urban—micro watersheds drained to the NBS.

NBS—time to deliver? During the last decade, EU countries have undertaken legal translation of the goals outlined in the Flood Directive's goals into their existing FR management. This process strives to influence policy (for example, by establishment of watershed jurisdictions) as well as practice (evidence-based decision making using cost cost-benefit analysis, collaborative management by public participation, enhancing ecological status by win-win measures in the light of Water Directive and more). These days, a substantial body of research is dedicated to comparative analysis of EU-countries, their respective FR governance setting and resulting policies and policy tools in the view of legal, planning, economics and participatory aspects (Raška 2015; Hegger et al. 2016b; Liefferink et al. 2018). On the contrary, little if any research is undertaken regarding the actual effectiveness of the watershed (catchment) during floods in of the watershed (catchment) FR management plans (the fruits of the governance and policy tools)—most probably because 11 years are not a sufficient time span for determinations of this kind. At this point in time, we must be able, at least to some extent, to formulate indicators that may predict effectiveness of established or planned to be established policy tools and measures—grey and green. Perceived effectiveness might be substituted by another term—legitimacy (Melnychuk 2017). For NBS concept and practice—is enough being done to deliver the legitimacy of the practice?

Feet on the ground (of a wetland)—Realistic and useful assessment of NBS. A decade ago, in 2009, a milestone work by Daily et al. stressed the failure of the ecosystem services concept to reincarnate into common practice. The Pilsen case study, as many NBS case studies, struggles with a similar pitfall—the turnover from a positive project into a relevant, accepted and eventually desired and common in practice. Nature-based solutions are measures aiming to relief relieve natural risks, in our case, flood risks, by enhancing and restoring the regulation of ecosystem services and are part of a sustainable FR management. Like ecosystem services, sustainable FR management outputs seek recognition and legitimacy (Melnychuk 2017). Not only should they be hydrologically sound, economically feasible, ecologically acceptable and publically supported—legitimate sustainable FR management also aims to meet local perception of floods and flood risks, echoes the prevailing socio-institutional factors and provokes lively discourse within the professional community (Goulden et al. 2018). The latest bon ton of relating to NBS as a superior strategy (Keesstra et al. 2018), while the actual relief, though plausible, is still evidence deprived and needs to be empirically scrutinized (Hegger et al. 2016a, b; Niazi et al. 2017) undermines the colossal multi-disciplinary efforts to provide NBS with a sound ground of evidence and hands-on legitimate experience.

Pilsen case—study in the light of FR governance capacity buildup. In the case of the City of Pilsen, Czech Republic, the small-scale NBS for frequent (though not in recent years) floods was a three-phase wetland area restoration in the urban floodplain of the Berounka river. The project aims for adequate prioritization of flood risks in spatial planning and public fund relocation for the rehabilitation of specifically designated urban "flood sponge" areas—flood water naturally retaining an area adjacent to the river bank. The first phase of the project was launched based on external (EU) funds and publicly owned land, and the subsequent phases (2 and 3) are dependent on the political will of local decision makers to spend public money for buying or long-term leasing privately owned land adjacent to phase 1.

- Challenges: Dominant public support favours grey infrastructure; the project has little impact on flood risk relief; a vast area of the planned project is privately owned; low level of political will—as decision makers are restrained from negotiating with private owners; policy makers are reluctant to undertaking undertake long-term trans-sectoral projects; their awareness of diverse measures was low; due to droughts, constructed wetlands did meet the groundwater.
- Opportunities: The landowners in the flood plain have limited opportunities for land use; the policy makers were reluctant to undertake vast upstream-downstream obligations and prefer to realize a local project.

From the mentioned public support for grey infrastructure and policymakers unaware of other possibilities, it seems that the prevailing capacity in Czech Republic's flood risk governance is resistance (see Table 13.1). Capacity to absorb and recover possibly exists as well, since the authors mention available spatial planning instruments in the form of zoning plans that prohibit future development in flood flood-prone areas—landowners near the river have limited opportunities for land use due to the floodplain, the Q100 active zone and the land-use plan. Described case study of wetland restoration belongs, as well to absorb and recover capacity. The described difficulties to communicate NBS benefits to policymakers imply low current capacity to transform and adapt.

The choice between one rationale (capacity) over another is inherently political, and the Pilsen case study describes the attempt of a group of "enablers" (the case description doesn't expand on their identity nor on their institutional affiliation) to engage local politicians and stakeholders in what seems like capacity diversification. Only sincere examination and acknowledgement of prevailing FR capacity and its basis will allow solid strategic decision-making regarding FR policy tools development. Weather the decision is to diversify the existing capacity, Table 13.2 presents three aspects, without which new concepts and tools will not be realized.

Existing effective and legitimate case studies expand the aspects of "knowing" as well as "wanting" and "enabling" dimension among the local state, municipality and political stakeholders. This process will add, shift and re-set some of the core assumptions underlying the "territorial intelligence" of the specific district/municipality,

Table 13.1 Different types of strategies in dealing with FR

Capacity type	Implication to FR management
Resistance	The ability to withstand disturbances—Mainly engineered structures. Dikes, Levees, etc.
Adsorb and recover	System is affected, but is still capable of responding and recovering—floodplains, NBS
Transform and adapt	Ability of a system and of institutions to adjust to climate change (including climate variability and extremes) to moderate potential damages, to take advantage of opportunities, cope with the consequences and learn

Based on Hegger et al. (2016a)

Table 13.2 Three "must –haves" for expanding governance capacity

Capacity type		Implication to FR management
Knowing	Actors need to be fully aware, understand, and learn the actual or possible risks and impacts of actions, policy, and strategic choices	• Increasing awareness and sense of urgency • Developing useful knowledge through available information • Cross-stakeholder learning
Wanting	Actors need to commit to cooperate, express, and act upon ambitions, and apply their skills and capabilities to find solutions	Engaging actors and ensuring their commitment to collaborate and truly contribute their skills to find solutions, sometimes with assistance of visionary agents
Enabling	Actors need to have the network, resources, and instruments to enable them to implement their ambitions	Defining authorities and responsibilities, developing and defining sources of reliable funding, collaborative policy development

Koop et al. (2017)

eventually expanding the capacities of FRM in a certain area (Kaufmann and Wiering 2017).

Nature-based solutions as part of a whole-puzzle: Structured "window" to the specific case studies and their status in the light of the flood risk governance ecosystem are in need and they should elaborate on the following suggested topics: (1) Translation of flood risk Directive to national laws; (2) Establishment of new or rearranging existing watershed jurisdictions; (3) Institutional setting and funding to comply with the FRM in the light of the historical or expected flood impact; (4) FR "reflection" through the territorial intelligence prism; (5) Relevancy of the NBS in the light of hazard maps and risk maps—all of the above will provide the necessary framing, and ultimately, legitimacy, to the case (Dieperink et al. 2016; Hegger et al. 2016a; Goulden et al. 2018; Kaufmann and Wiering 2017; Koop et al. 2017, 2018; Nordin von Platen and Gustafsson 2018; Wiering et al. 2018). A solid example for such a type of "structured" window would be the study of Hedelin (2016). Further theoretical examination of the "FR chain" in different states, different watershed types and climates will allow for the development of systematic evaluation tools according to the FR ecosystem typologies.

First things first—who is in charge? Studies in some EU countries show that, while the central state remained the core actor in flood risk governance, the decentralization process transferred significant responsibilities to the municipalities (Wiering et al. 2018; Liefferink et al. 2018). In other EU countries, such as Ireland, the National Office of Public Works (OPW) is the jurisdiction to map the risks and to run the FRM plans. A clear depiction of the players in charge sheds light on the type of constrains they might experience, which should be specifically addressed.

What flood risk atmosphere supports NBS? In the light of frequent floods in Pilsen of the past, the decision makers' survey showed that floods are the most important hazard in the city. Nevertheless, the new strategic plan for Pilsen lacks any reference to flood risks. These two findings suggest that better communication of high flood risk to the planners, and other decision makers is in need. Though higher risk perceptions increase flood preparedness intentions, nothing assures that this preparedness will result in diversification in management measures. Once the flood risks is perceived as significant, and the prevailing FRM strategy is resistance, 'NBS' are a rather unexpected outcome. The question arises,—how to rise flood risk importance, but in a 'NBS'—supportive way? What is the plausible 'flood atmosphere' so NBS are considered widely? Two strategies come into mind: (1) NBS as a tailor-made, win-win solution, supported by strong hydrological evidences, on the ground of a detailed, informative, clear watershed FRM plan. (2) NBS as a "better option" for parks in the city, which also act as some sort of relief during floods.

How ecological is the natural? Though the term "ecology" is abundant in the NBS agenda, it nearly never stands alone and is more often than not linked to economics ("ecological economics"), engineering ("ecological engineering") and humans as in "socio-ecological aspects" (Kabisch et al. 2016). Resilience and liveability of the cities, presumably resulting from blue-green infrastructure, nearly never raise a discourse on the core-ecological dimensions of the NBS. Wetlands in general are poorly understood habitats. Though quite a few NBS, like those in the Pilsen case, rely on wetland runoff retention capacity, its reclamation efforts rarely aim for "a good ecological status" of the NBS. How important is the ecological state to the NBS's function? What is the desired ecological state of an NBS relying on wetlands, meanders and river restoration?

NBS are about the process. The appropriateness of a particular NBS in a certain geographical, social, administrative and cultural context is of high importance. But even for the most appropriate project, the transaction costs are high—mapping the institutional setting and stakeholders' interests and barriers, revealing attitudes regarding floods, setting steering committees, communicating the risks and the various possible measures, setting the financial grounds—it requires time, a strong will and patience.

Mapping internal and external barriers for NBS implementation: NBS marginalisation as a meaningful solution to FR is explained, in the case of Pilsen, by how unaware stakeholders were of the importance of NBS and their benefits.

This statement falls in line with the Linear Model of pro-environmental behaviour, prevailing during the 70s and 80s of the previous century, implying direct correlation between knowing and doing (Ervin and Ervin 1982). This model was proven to be insufficient by a massive body of research on barriers for environmental behaviour. Stakeholders' setbacks can vary from internal (such as hate of risk, past personal experience) to external (such as institutional setting, lack of funding opportunities, etc.) (Kollmuss and Agyeman 2002). In the field of FR management research, mapping barriers is still rare although it can contribute vastly to a clearer understanding of the governance challenges (Dieperink et al. 2016; Goulden et al. 2018).

Strong pressure by specific interests (actor coalitions) is a major force of change in flood risk governance (Liefferink et al. 2018). Nature-based solutions are advocated for and promoted by different authorities, from the federal and/or state level, NGO's, academia, local municipal "champions" or a semi-structured arrangement of all of the above-mentioned, together forming a bricolage of flood risk governance (Merrey and Cook 2012).

In the Pilsen case, the decision makers were reluctant to engage in the negotiation process with private landowners. Since phase 1 of the wetland restoration didn't involve such negotiations, this capacity might be lacking and should be strengthened, or the part of negotiation could be performed by other actors—suggesting a possible intermediator role for a local NGO, similar to the case of the Thau Watershed (Etang de Thau), where all types of negotiations were led by animators—knowledge brokers between top-down public policies (including water policies) and bottom up development projects—supported by local population and politicians (Bouleau et al. 2009). The legitimacy of animateurs stems from the network they build around state offices, water professionals, local authorities and their services, elected politicians, NGOs, schools, journalists and the larger public. Animateurs gain their legitimacy from their interpersonal communication capacities, from their function as knowledge hubs and how they foster and support participation and commitment among the experts and stakeholders participation and commitment to within their networks (Plant et al. 2014).

Seize opportunities to communicate flood risks perception. Studies show that the level of interest in flood hazard and risk generally decreases between the individual events and that the flood risk is not perceived as a substantial and permanent characteristic of the local environment, but rather in terms of discrete flood events. This can also lead to perceiving the flood as less of a threat over time (Raška 2015). In Pilsen, it appears that even though the last years were characterized by drought rather than floods, decision makers agreed on the high degree of importance of floods and flood managing projects. In Pilsen, floodplains and other areas are under development restrictions (Q100 active zone) in the land-use plan of the city.

Complexity is no excuse—and we need more theory to guide us. Once developing policy in the light of its implementation, in an era when governance is beyond official title, policy makers as well as academia have little excuses to turn a blind eye to the realm of the FRM complexity. The seeming line between the physical inventories of the Watershed to its human dimensions, in the view of collaborative policy making, is stunningly complex. Utilising theory makes it easier to embrace this complexity and its dimensions (Hedelin 2018). This calls for complementary theory-based approaches that allow for policies with a widened scope of principles (Kaufmann and Wiering 2017). This is especially true for NBS legitimacy acquisition as an FR easement.

Acknowledgements Open access of this chapter is funded by COST Action No. CA16209 Natural flood retention on private land, LAND4FLOOD (www.land4flood.eu), supported by COST (European Cooperation in Science and Technology).

References

Bouleau G, Barone S, Maurel P, Abrami G, Cernesson F, Richard S (2009) I-five: innovative instruments and institutions in implementing the water framework directive French case study report: implementing the WFD on the Thau basin. www.gwp.org. Accessed 15 Mar 2019

Brunckhorst D, Reeve I (2006) A geography of place: principles and application for defining 'eco-civic' resource governance regions. Aust Geogr 37(2):147–166

Daily GC, Polasky S, Goldstein J, Kareiva PM, Mooney HA, Pejchar L, Ricketts TH, Salzman J, Shallenberger R (2009) Ecosystem services in decision making: time to deliver. Front Ecol Environ 7(1):21–28

Dieperink C, Hegger DT, Bakker MHN, Kundzewicz ZW, Green C, Driessen PPJ (2016) Recurrent governance challenges in the implementation and alignment of flood risk management strategies: a review. Water Resour Manage 30(13):4467–4481

Ervin CA, Ervin DE (1982) Factors affecting the use of soil conservation practices: hypotheses, evidence, and policy implications. Land Econ 58(3):277–292

Goulden S, Portman ME, Carmon N, Alon-Mozes T (2018) From conventional drainage to sustainable stormwater management: Beyond the technical challenges. J Environ Manag 219:37–45

Hedelin B (2016) The EU Floods directive trickling down: tracing the ideas of integrated and participatory flood risk management in Sweden. Water Policy 19(2):286–303

Hedelin B (2018) Complexity is no excuse. Sustain Sci 1–17

Hegger DL, Driessen PP, Wiering M, Van Rijswick HF, Kundzewicz ZW, Matczak P, Crabbé A, Raadgever GT, Bakker MH, Priest SJ, Larrue C (2016a) Toward more flood resilience: is a diversification of flood risk management strategies the way forward? Ecol Soc 21(4):53

Hegger DLT, Driessen PPJ, Bakker M, Alexander M, Beyers JC, Buijze AWGJ, Chorynski A, Crabbé A, Deketelaere K, Delvaux B, Dieperink C (2016b) A view on more resilient flood risk governance: key conclusions of the STAR-FLOOD project. STAR-FLOOD Consortium

Kabisch N, Frantzeskaki N, Pauleit S, Naumann S, Davis M, Artmann M, Haase D, Knapp S, Korn H, Stadler J, Zaunberger K (2016) Nature-based solutions to climate change mitigation and adaptation in urban areas: perspectives on indicators, knowledge gaps, barriers, and opportunities for action. Ecol Soc 21(2):39

Kaufmann M, Wiering M (2017) Discursive junctions in flood risk governance—a comparative understanding in six European countries. J Environ Manag 196:376–386

Keesstra S, Nunes J, Novara A, Finger D, Avelar D, Kalantari Z, Cerdà A (2018) The superior effect of nature based solutions in land management for enhancing ecosystem services. Sci Total Environ 610:997–1009

Kollmuss A, Agyeman J (2002) Mind the gap: why do people act environmentally and what are the barriers to pro-environmental behavior? Environ Educ Res 8(3):239–260

Koop SHA, Koetsier L, Doornhof A, Reinstra O, Van Leeuwen CJ, Brouwer S, Dieperink C, Driessen PPJ (2017) Assessing the governance capacity of cities to address challenges of water, waste, and climate change. Water Resour Manage 31(11):3427–3443

Koop S, Monteiro Gomes F, Schoot L, Dieperink C, Driessen P, Van Leeuwen K, (2018) Assessing the capacity to govern flood risk in cities and the role of contextual factors. Sustainability 10(8): 2869

Liefferink D, Wiering M, Crabbé A, Hegger D (2018) Explaining stability and change. Comparing flood risk governance in Belgium, France, the Netherlands, and Poland. J Flood Risk Manag 11(3): 281–290

Melnychuk N (2017) Assessing legitimacy within collaborative water governance: how, when, and by whom? PhD dissertation, University of Waterloo

Merrey DJ, Cook SE (2012) Fostering institutional creativity at multiple levels: towards facilitated institutional bricolage. Water Altern 5(1):1–19

Niazi M, Nietch C, Maghrebi M, Jackson N, Bennett BR, Tryby M, Massoudieh A (2017) Storm water management model: performance review and gap analysis. J Sustain Water Built Environ 3(2):1–32

Nordin von Platen H, Gustafsson M (2018) Nature-based solutions for flood risk reduction, contamination control and climate change adaption

Plant R, Maurel P, Barreteau O, Bertacchini Y (2014) The role of territorial intelligence: the case of the Thau territory, Southern France. River basin management in the twenty-first century: understanding people and place, 446–466

Raška P (2015) Flood risk perception in Central-Eastern European members states of the EU: a review. Nat Hazards 79(3):2163–2179

Rittel H, Webber M (1973) Dilemmas in a general theory of planning. Policy Sci 4(2):155–169

van Ruiten L, Hartmann T (2016) The spatial turn and the scenario approach in flood risk management: implementing the European floods directive in the Netherlands. AIMS Environ Sci 3(4):697–713

Wiering M, Liefferink D, Crabbé A (2018) Stability and change in flood risk governance: on path dependencies and change agents. J Flood Risk Manag 11(3):230–238

Jenia Gutman works for the Israeli Ministry of Agriculture and Rural Development, in the Drainage and Soil Conservation Department, coordinating strategic intra-ministerial policy making processes: Development of Flood Risk Management Policy in Israel (EU Flood Directive adoption) according to three main strategies—prevention, protection and mitigation. Additionally, she coordinates the Israelian integrated watershed management policy steering committee and works on tool development and assimilation, together with local champions, stakeholders and academics. Farmland and local agriculture conservation, in the light of its multi-functionality, is her main focus of interest, together with international best practice assimilation.

Chapter 14
Commentary: Flood Retention in Urban Floodplains—A Plzen Case Study from the Viewpoint of a Hydraulic Engineer

Reinhard Pohl

Introduction

Since ancient times, people have been settling near rivers that were used for water supply, irrigation, wastewater disposal and. in the case of larger rivers, as a navigable waterway, as a border or line of defense. On the one hand, the suspended matters of the rivers fertilized the temporarily inundated land during floods, and, on the other hand, floods could endanger the people living in the flood-prone area. Thus, the two main issues of hydraulic engineering (building structures for the use of water and protection against the water) remain an everlasting challenge.

As natural rivers are wide, meandering and not very deep, they were often accompanied by large wetlands and bends that were able to store water during floods. Inundations of these areas were a normal process without hazards and damages because these can only occur in a man-made environment. Later the river-near land was needed and used for settlement and economic activities so that an inundation could cause damages and the area had to be flood-protected. Reducing the inundation area could cause a reduction of the retention effect so that floods with less reduced peaks were able to pass.

The Lobezská Louka Site

The rivers Úhlava, **Úslava** (65.7 km), Radbuza, and Mže have their confluence in or near Plzen (Pilsen) to form the river Berounka, which is a left tributary of the Vltava (Moldau) with the junction upstream of the city of Prague. The Lobezská Louka (meadow) area at the Úslava river in the city of Plzen was an unmanaged green

---

R. Pohl (✉)
TU Dresden/Institut für Wasserbau und Technische Hydromechanik, Dresden, Germany
e-mail: Reinhard.Pohl@TU-Dresden.de

T. Hartmann et al. (eds.), *Nature-Based Flood Risk Management on Private Land*,
https://doi.org/10.1007/978-3-030-23842-1_14

overgrown with wild tree seedlings and with occasional illegal dumps. The area of the original wetlands was split into lots with several owners, who were following different interests with their estates.

In the addressed chapter, the authors Macháč and Louda report on this area in Plzen that was given back to the Úslava river. By doing so, wetlands with a nature-near state were restored to improve the possibilities for recreation, biological diversity and flood protection.

Hydraulic Aspects of Wetland Retention

In order to understand flood routing and protection, one must know that the flow in every river or channel is subjected to a small peak reduction, which is caused by the different hydraulic gradients of the coming and leaving flood wave (Bornschein and Pohl 2017). This brings a higher propagation celerity of the flood wavefront and a slower velocity on the reverse side of the wave. If a considerable cut of the peak is desired or needed, large storage volumes are required that should be gated additionally so that the storage is not filled before the arrival of the flood peak.

In the presented case, this means that the estimated mean flow of 1.5 m^3/s can be stored for 1–2 h assuming a storage of 8000 m^3. A 100-year-flood with a peak of more than 200 m^3/s will not be retarded or reduced by the recreated wetlands because the flood wave volume of some million cubic meters will fill the storage area completely within only a few minutes, long before the peak arrives. The above numbers are only estimations because they were not given in the original paper and not available from the online catchment basin information (rivers.raft.cz/cechy/uslava.aspx). Considering the relative large catchment area of more than 90 km^2, they might be still underestimated.

Figure 14.1 shows example hydrographs of a small flood (say 2-yr flood) and a large flood (say 100-yr flood) in comparison. All areas in the chart indicate water volumes because the vertical diagram axis represents the discharge and the horizontal axis depicts the time. It is visible that, for the small event, a peak reduction is possible, whereas, for the large event, a controlled storage is required. Only when its inlet gates are opened at point P the expected peak reduction can be reached.

Often it is postulated that a couple of small protection measures are better than one large measure. As the economic and hydraulic efficiency depends on many influences, it cannot be said without profound analysis of each particular case whether one large or several smaller flood protection measures will bring the better effect. However, it can be said that every measure to revitalize an industrial wasteland or not used urban "grey" area will be a facelift for the city as well as a step towards a nature-near river situation.

In this analysis, many input values have to be estimated or adopted from other case studies. From this fact, a considerable uncertainty of the results arises. Another issue is the comparison of values with different dimensions because not all relevant quantities or qualitative properties can be converted to monetary values that can be

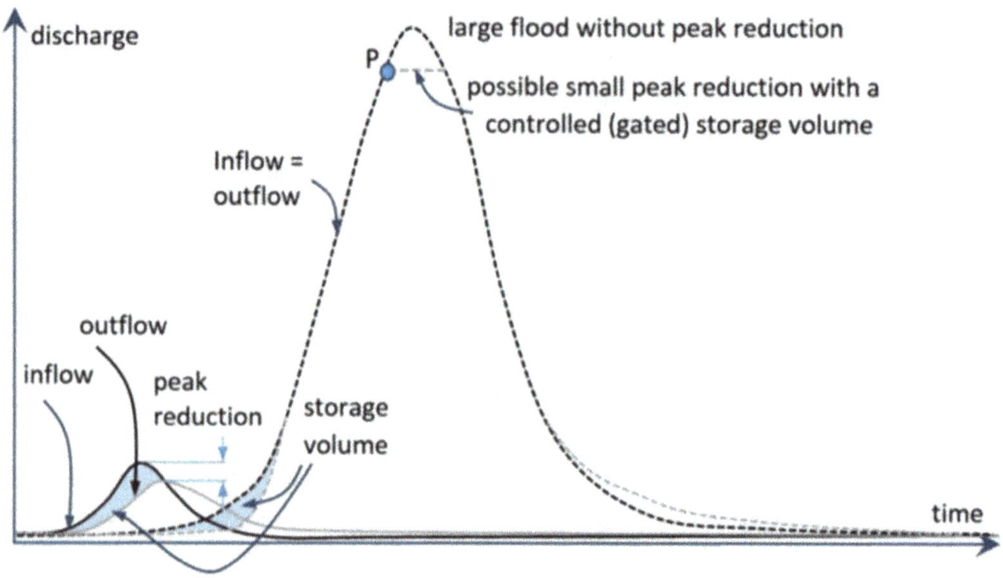

Fig. 14.1 Effect of a small nature-based flood protection measure. Comparison of a small flood (solid lines) and a large flood (dotted lines). Inflow hydrograph: black line, outflow: grey lines. Hatched area: storage volume

expressed in terms of a currency. In such cases, a Pareto optimization may help to indicate improving items without worsening of other items.

When speaking about NBS or non-structural methods, we must also consider that these projects need a lot of construction work, at least during the phase of project implementation, but in many cases also during its lifetime. Insofar these measures are structural measures including earthwork, excavation, reinforcement of embankments, building pathways and roads and in some cases also bridges, inlet/outlet structures, and flood defences.

Conclusion

Small measures can only affect small floods. Nevertheless, they can help to reduce the frequency of inundations downstream. In the present case with a retention storage capacity of about 8000 m^3, only protection against very frequent events might be expected. When calculating costs and benefits, the main benefit in the Lobezská Louka (meadow) area might not arise from the flood protection but from other effects like restoration of the floodplain area, removal of illegal rubbish dumps, establishing a recreation area for inhabitants, etc. (Dittmann et al. 2009).

The example of the Lobezka Louka demonstrates that it is possible to recreate an urban area to become a wetland close to the assumed original situation before the urban development. At this juncture, the cooperation of all stakeholders, including the landowners, is essential. This will allow land recycling instead of additional land consumption.

Acknowledgements Open access of this chapter is funded by COST Action No. CA16209 Natural flood retention on private land, LAND4FLOOD (www.land4flood.eu), supported by COST (European Cooperation in Science and Technology).

References

Bornschein A, Pohl R (2017) Land use influence on flood routing and retention from the viewpoint of hydromechanics. J Flood Risk Manag 403(2011):103. https://doi.org/10.1111/jfr3.12289

Dittmann R, Froehlich F, Pohl R, Ostrowski M (2009) Optimum multi-objective reservoir operation with emphasis on flood control and ecology. Nat Hazards Earth Syst Sci NHESS 9(6):1973–1980

Reinhard Pohl studied Hydraulic Engineering and got his Ph.D. from the Dresden University of Technology. After four years with Hydroprojekt Consultants, he came back to the University in 1993. His working fields are hydromechanics, flood protection, dams, weirs, levees and the safety of hydraulic structures. After receiving the venia legendi in 1998, he became a Professor of hydromechanics in civil engineering in 2002. He is author or co-author of several textbooks, German DIN standards and guidelines on dams and levees and many scientific papers.

Part VI
A Regional Flood Management in Austria

Chapter 15
Blauzone Rheintal: A Regional Planning Instrument for Future-Oriented Flood Management in a Dynamic Risk Environment

Lukas Löschner, Walter Seher, Ralf Nordbeck and Manfred Kopf

This chapter explores the case of a regulatory spatial planning instrument for flood risk management (FRM). The so-called "Blauzone Rheintal", a regional plan designating large-scale areas for flood retention and flood runoff in the Austrian Rhine Valley, was issued in 2013 by the Vorarlberg state government to secure flood hazard areas and mitigate future increases in flood risk. The selected case highlights the potential for spatial planning to support nature-based solutions (NBS) in FRM, particularly to secure land resources for the implementation of (land-intensive) flood retention measures. Based on a document analysis and qualitative interviews with the leading experts of the "Blauzone Rheintal" and with planning authorities in the Swiss canton St. Gallen, this contribution unravels the processes leading up to the final spatial plan, including the cross-sectoral and multilevel interactions that balanced the conflicting (land use) interests in the study area. Given the long-term dimension of

L. Löschner (✉) · W. Seher
Institute of Spatial Planning, Environmental Planning and Land Rearrangement (IRUB), Vienna, Austria
e-mail: lukas.loeschner@boku.ac.at

W. Seher
e-mail: walter.seher@boku.ac.at

Department of Landscape, Spatial and Infrastructure Sciences (RALI), Vienna, Austria

L. Löschner · W. Seher · R. Nordbeck
University of Natural Resources and Life Sciences Vienna (BOKU), Vienna, Austria
e-mail: ralf.nordbeck@boku.ac.at

R. Nordbeck
Institute of Forest, Environmental, and Natural Resource Policy (InFER), Vienna, Austria

Department of Economics and Social Sciences (WiSo), Vienna, Austria

M. Kopf
Spatial Planning Department, State Government of Vorarlberg, Bregenz, Austria
e-mail: manfred.kopf@vorarlberg.at

T. Hartmann et al. (eds.), *Nature-Based Flood Risk Management on Private Land*,
https://doi.org/10.1007/978-3-030-23842-1_15

many NBS, this contribution, moreover, discusses the apparent contradiction between static planning instruments and the need for adaptive, forward-looking FRM.

Introduction

Due to the concentration of inhabitants, settlements and economic activities in valleys, alpine areas are particularly susceptible to the impacts of natural hazards (Petraschek and Kienholz 2003; Fuchs et al. 2015). Since the mid-1990s, areas north of the alpine ridge have experienced a series of devastating flood events (1999, 2002, 2005 and 2013). Heavy losses, especially in densely populated alpine valleys, made evident the limits of traditional, defence-oriented flood policies and highlighted the need for more integrated approaches to FRM based on a portfolio of structural and non-structural measures (BMLFUW 2009).

A fundamental principle in the nascent policy paradigm is to "make space for water" (Warner et al. 2012). Across Europe, flood managers are increasingly reconnecting floodplains and providing more room for flood runoff and water retention to alleviate flood risks as well as to enhance the ecological and recreational values of riverscapes (Samuels et al. 2006; Klijn et al. 2008). Especially in topographically confined alpine regions, however, land is a limiting factor for the implementation of non-structural flood risk reduction measures. In these areas, spatial planning is of particular relevance to better allocate space to competing demands (Hartmann 2011).

This chapter explores the case of a regulatory spatial planning instrument for FRM. The so-called "Blauzone Rheintal", a regional plan designating large-scale areas for flood retention and flood runoff in the Austrian Rhine Valley, was issued in 2013 by the Vorarlberg state government to secure flood hazard areas and mitigate future increases in flood risk (State Government of Vorarlberg 2013). The selected case highlights the potential for spatial planning to support NBS in FRM, particularly to secure land resources for the implementation of land-intensive flood retention measures. Based on a review of policy documents in water management and spatial planning as well as four semi-structured interviews with the leading experts of the "Blauzone Rheintal", the case study unravels the process leading up to the final spatial plan, including the cross-sectoral interactions and the engagement of stakeholder groups to balance conflicting land use interests in the study area. Given the long-term dimension of many NBS, this contribution moreover discusses the apparent contradiction between static planning instruments and the need for adaptive, forward-looking FRM.

The chapter is organized as follows: the following section briefly outlines the geographic and institutional context of the "Blauzone Rheintal"; the section thereafter provides a detailed description of the planning instrument and the associated restrictions in land use and property rights; the subsequent section traces the policy process leading up to the spatial plan, including intersectoral coordination between water management and spatial planning and the involvement of different stakeholder groups; the final section discusses the relation of spatial planning and NBS and draws conclusions regarding the possibility of replicating and scaling-up the "Blauzone" in other flood-prone regions.

The Rhine Valley: A Dynamic and Vulnerable Region in the Heart of Europe

The Rhine Valley is an alpine valley in Central Europe that extends over 90 km along the Rhine from the source of the Rhine in Switzerland via Liechtenstein to Austria and Germany. The lower part of the Rhine Valley, separated by the Rhine into Austrian and Swiss areas, forms a wide basin ranging from the alpine ridge to the Bodensee (see Fig. 15.1).

Given its favourable geographic location and topographic conditions for agricultural production, this part of the Rhine Valley is a historical settlement area that today ranks as one of the most dynamic regions in Central Europe. In particular, the Austrian side of the Rhine Valley (Vorarlberger Rheintal) is experiencing strong settlement growth and economic development. Since 2006, the residential population has increased by 8.5% (to approximately 260,000, or two thirds of the total population of Vorarlberg), and it is expected that the bulk of the province's future population growth (+16% until 2050) will be concentrated in the Rhine Valley (State

Fig. 15.1 View of the Rhine Valley with the Rhine outlet in the Bodensee and the Alps in the background. The Rhine forms the border between the Austrian part (left) and the Swiss part of the Rhine Valley [adapted from Krapf 2012; Wikipedia/Creative Commons]

Government of Vorarlberg 2016a). This also reflects the region's strong economic relevance: As one of the Austria's leading manufacturing regions, the Rhine Valley accounts for more than 70% of economic production in Vorarlberg (WKO 2017).

Due to the concentration of vulnerable land uses, the region is at a high risk of flooding (BfG 2015; BMLFUW 2015). Currently, areas along the alpine part of the Rhine (Alpenrhein) is protected against the impacts of floods with a 100-year return period (discharge of 3.100 m^3/s), but simulations show that low-probability events (i.e., floods with a 300-year return period) would produce economic losses of around EUR 2.7 billion in Austria and EUR 2.07 billion in Switzerland (IRR 2017). To reduce the risk resulting from such an extreme flooding scenario, in 2005, the state governments of Vorarlberg, St. Gallen (CH), Graubünden (CH) and Liechtenstein initiated the transnational flood protection program "Rhesi" with the principal aim of raising the flood protection standard to 4.300 m^3/s (i.e., the 300-year discharge level), providing emergency retention areas and preventing uncontrolled dam breaches (IRR 2017).

The transnational character of the Rhesi project, however, is not matched by a common regulatory planning approach because spatial planning is widely regarded as a national task. In Austria, the spatial planning system is a comprehensive system arranged in a hierarchy, where the states are in charge of legislation and policy-making. The implementation of spatial planning is divided between the states and the municipalities, and the states are responsible for planning issues of regional importance. Local planning issues are controlled by the municipalities. In spatial planning practice, the local planning level (where the local land use plan is the most important instrument) is by far more relevant than the regional level. In comparison to other federally organized countries such as Germany or Switzerland, regional planning in Austria generally has a weak position. Due to the strong constitutional and political autonomy of Austrian municipalities, state governments have limited scope to interfere in municipal land use planning through regulatory instruments (Marik 2005).

The comparably low importance ascribed to regional planning applies particularly to flood-related spatial planning. With the exception of the "Blauzone" directive, Austrian spatial planning laws to date provide no legal obligations concerning direct flood-related zoning in regional planning, notwithstanding the wide availability of flood hazard information on a regional scale. By preserving open areas for flood retention and higher discharge levels and securing areas for future flood control measures, the spatial plan "Blauzone Rheintal" supports the implementation of the Rhesi project. This underpins the unique position of the *Blauzone*, giving this approach a role model function in Austrian spatial planning. The *Blauzone* also stands out in a cross-border perspective as comparable regulatory plans are missing on the Swiss side of the Rhine (Canton St. Gallen), despite a strong role of the cantonal level in spatial planning. Planning authorities in St. Gallen, however, highlight, that the cantonal structure plan (amended in 2017) legally defines the long-term settlement boundaries to prevent urban sprawl into agricultural land and thus preserves potential retention areas for emergency flood relief (Interview 4).

Blauzone Rheintal: A Regional Plan to Secure Large-Scale Flood Areas

In 2005, Alpine regions in Austria and Switzerland suffered a major flood event. Vorarlberg, the westernmost province in Austria, was particularly affected, as some regions experienced the highest discharge levels in more than a century. Total damages to households, businesses, infrastructure, etc., amounted to more than EUR 180 million, making it the costliest natural disaster in Vorarlberg to date (State Government of Vorarlberg 2015). The majority of the flood damage was recorded in the province's alpine valleys, whereas the densely populated Rhine Valley with its concentration of economic assets was affected only to a limited extent. In the ten years following the seminal flood events (2005–2015), more than EUR 300 million was invested in flood protection, and an additional EUR 200 million has been provided until 2020 (State Government of Vorarlberg 2016b). Many of these flood protection measures were implemented or are planned in areas with strong settlement pressure, such as the Rhine Valley.

One of the key challenges for flood policy makers following the flood events in 2005 was providing space for flood alleviation measures and preventing urban sprawl into potential hazard areas. Faced with a lack of appropriate regulatory instruments to secure large-scale areas for flood protection measures, state officials in water management and spatial planning engaged in an intersectoral coordination process to identify and delineate suitable areas. In 2013, following another large flood event, the Vorarlberg state government issued the "Blauzone Rheintal", a legally binding regional spatial plan that designates flood runoff and flood retention areas along the Rhine and its tributaries (see Fig. 15.2). The planning instrument pursues the following aims (State Government of Vorarlberg 2013):

- *Protect settlement areas*: Existing settlement areas (i.e., built areas as well areas zoned as building land in the local land use plans) are protected against flooding. To minimize the further increase in damage potential, zoning for building land in the designated flood hazard areas is severely restricted.
- *Preserve open areas for flood retention and flood discharge*: To reduce flood peaks, existing and potential flood retention areas are kept free of building development. These areas particularly include agricultural and forest areas with low damage potential, which may also be temporarily flooded in extreme events, as when there is a need to accommodate storm water to prevent a dike breach.
- *Secure areas for future flood control measures*: Areas with low damage potential are secured for future flood control measures in order to preserve long-term FRM options.

Covering an area of 5.400 ha in twenty-two municipalities on the Austrian side of the Rhine Valley, the "Blauzone" is overwhelmingly (ca. 90%) located in the so-called "Grünzone", which was established in 1977 to preserve open spaces and agricultural areas of regional importance in the Rhine Valley. Accordingly, the "Blauzone" predominately includes areas with low damage potential, such as agricultural

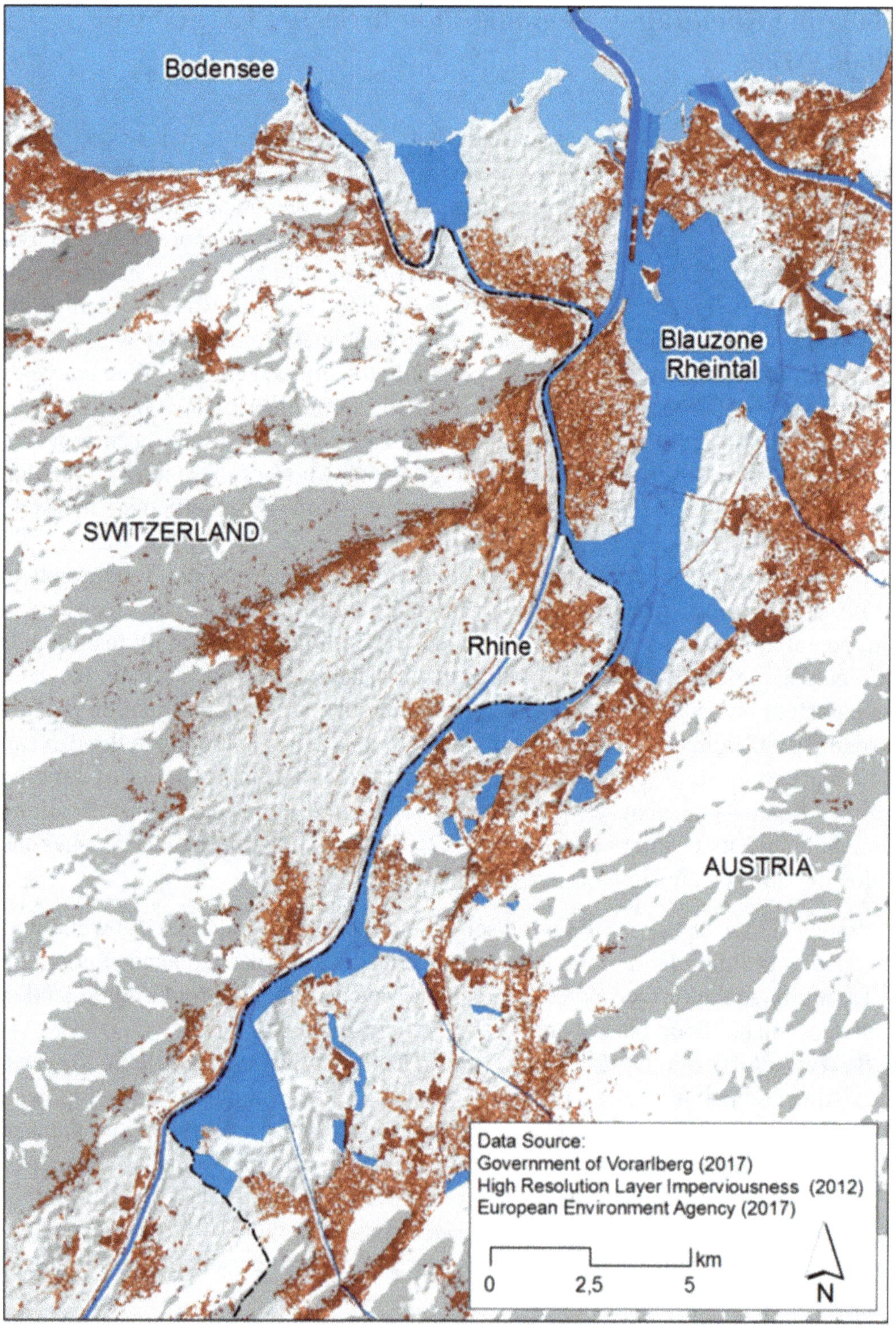

Fig. 15.2 Spatial extent of the Blauzone Rheintal

or forestland. Highly vulnerable areas, meaning developed areas, as well as land zoned for building were specifically excluded from the spatial plan.

Given its character as a regulatory planning instrument, the "Blauzone" was issued as a legally binding directive, which obliges the affected municipalities to amend their local land use plans and (re)zone areas located within the "Blauzone" as so-called open space reserve areas. This means that no development is permitted in those reserve areas with the exception of enlarging existing agricultural facilities (State Government of Vorarlberg 2013).

Interests and Interactions: The Process Leading Up to the Regional Plan

Plans to establish a set of retention areas in the Rhine valley first emerged in 2005 from discussions in the context of the Development Concept Rhine Valley (IRKA and IRR 2005). The development concept was adopted in December 2005 and introduced the idea of establishing runoff areas for extreme events along the Rhine. It proposed the formulation of overflow sections and runoff areas for the region and urged that these runoff areas be secured by spatial planning instruments. In May 2008, the Vorarlberg state government adopted various measures to coordinate spatial planning and flood protection in order to improve the spatial prerequisites for FRM and to secure runoff and retention areas as well as areas for future flood control measures. One of the suggested measures was the establishment of so-called "Blue Zones".

In the following years, the state authorities, in particular the departments of spatial planning and water management, continued to pursue the idea of "Blue Zones" (Frei and Kopf 2011). However, the final push to realize the regional plan came from the extreme flood event that occurred in June 2013 (Interview 1, Interview 2). Six months later, the regulation for the Blauzone Rheintal was passed by Vorarlberg's state parliament (State Government of Vorarlberg 2013). Subsequently, the regional plan was made available online via the state's geographic information portal (VOGIS).

During the planning process of the Blauzone, different user interests had to be taken into account and weighed to ensure that the public interest, as a higher-ranking goal of spatial planning, is served as well as possible. To accommodate the various interest groups, two elements of the planning process are important and will be analysed in greater detail: the intersectoral coordination between the spatial planning and water management authorities and the engagement of different stakeholder groups, particularly the municipalities, organized interest groups and affected landowners.

Intersectoral Coordination: The New Paradigm of Integrated Flood Risk Management in Action

Sectoral integration is defined as the integration of different public policy domains and includes the integration of public, private, and voluntary sector activities (Kidd 2007). Therefore, sectoral integration may be understood as a combination of sectoral policy integration and sectoral actors' integration. In practice, integration often refers to an approach to strengthen the linkages between places, the cooperation between sectors, or the interconnections among policies (Ran and Nedovic-Budic 2016). More integrated sectoral policies can encourage greater understanding of the effects of policies on other sectors. It can further help to promote synergies and consistency between policies in different sectors, improve the achievement of cross-cutting goals or objectives, and promote innovation in policy development and implementation (Stead and Meijers 2009).

Intersectoral coordination between flood management and spatial planning was identified as a major concern in Austria after the extreme flood events in 2002 and 2005. Several proposals have been made, inter alia, to create legally binding provisions for floodplains and hazard zones in spatial planning laws and building laws in order to prohibit the development of land that is important for flood runoff or retention (Habersack et al. 2009). Spatial planning is expected to contribute to flood mitigation mainly because it can influence the incidence of flooding and its consequential damage by regulating the locations of activities, types of land use, scales of development, and designs of physical structures (Ran and Nedovic-Budic 2016).

In response to the extreme flood events in 2005, policy makers in Vorarlberg formulated a strategy for integrated flood protection. The objective of the strategy is to minimize risks and improve the quality of overall flood protection. Based on the guiding principle of modern hydraulic engineering ("flood retention instead of flood acceleration"), the strategy specifically aims to preserve natural runoff areas, widen waterways and establish retention areas (State Government of Vorarlberg 2016b). To this end, the strategy highlights the need for better intersectoral coordination, particularly between water management and spatial planning, and defines spatial provisions as a strategic cornerstone for securing flood hazard areas and preventing the increase in potential damage in flood-prone areas.

The Blauzone Rheintal is a result of the strategic reorientation of the state's flood policy. It was developed on the basis of intersectoral coordination between the departments of spatial planning and water management and, thus, illustrates some of the challenges related to policy integration. Importantly, the Blauzone highlights that intersectoral policy-making is about balancing different policy interests. The spatial plan in its present form, in fact, represents a compromise between water management's sectoral demand for "more space for the river" and spatial planning's need to equally consider other public interests in the Rhine Valley, notably future opportunities for socio-economic development.

The intersectoral coordination processes leading up the delineation of the "Blauzone" may be characterized as a process of "spatial translation", which evolved around two issues: (i) data abstraction and harmonization and (ii) exclusion and contextualization. For one, the hydrological data and models for 300-year flood events and emergency runoff generated disconnected flooding areas, which could not be mapped into a single spatial plan. The data had to be transformed by the spatial planning authorities into a more coherent map of flood areas based on defining landscape and geographic features, such as roads, property lines or terrain edges (Interview 1). Second, the flood models had to be contextualized to facilitate implementation. All vulnerable land uses (i.e., all developed areas as well as land designated for building) were therefore eliminated from the flood models to minimize conflicts with landowners. Moreover, priority areas for future settlement development that were displayed in the spatial development plans were excluded to maintain socio-economic options for the municipalities (State Government of Vorarlberg 2013).

In summary, the integration of spatial planning with flood-risk management in the case of the Blauzone was based on three factors:

(i) Building communication channels between the departments and their representatives to develop a common mind-set based on collective knowledge and shared interests.
(ii) A coordinated management of information using geographic information systems as facilitators to support both spatial planning and flood-risk management.
(iii) A rational decision-making process that evaluated policy alternatives from both perspectives: the flood-risk management perspective as well as the planning and development perspective.

Information and Negotiation: Engaging Municipalities and Landowners

The "Blauzone" directive obliges a total of twenty-two municipalities to implement the 'Blue Zones' in local land use plans by (re)zoning areas as open space. To ensure a broad acceptance of the planning instrument, representatives of all affected municipalities as well as representatives of agricultural and commercial associations were informed about the spatial provisions in a series of events by staff from both the water management and the spatial planning authorities (Frei and Kopf 2011; Kopf 2016). Landowners and affected parties had the opportunity to comment on the proposed plans in a consultation procedure (Kopf 2016). As the spatial provisions overwhelmingly concern agricultural land, farmers and the Chamber of Agriculture in particular were strategic partners for the implementation of the regional plan.

On the one side, with the political backing of the state government, it was necessary to convince the representatives of the affected municipalities of the need to preserve open areas for flood retention and flood runoff and to secure areas for flood protection measures in the future. State representatives conducted information events with the local representatives of the 22 affected municipalities. This process of informing and convincing the communities was described as time-consuming and challenging

(Interview 2), and there were, of course, mayors who were more accessible from the beginning and others who were less so. Nevertheless, reservations on the part of the local representatives remained rather low, mainly because settlement areas and areas designated as building land were excluded from the "Blauzone" and the municipalities' development opportunities were hardly limited. On the other side, more work had to be performed to convince the affected landowners (in particular farmers) who acknowledged, on the one hand, that the open areas would be even more stringently protected by the "Blauzone" but, on the other hand, feared a reduction in land value and a restriction of future farm extensions (Interview 1, Interview 2). These reservations were countered by the legal possibility of attaching further buildings to an existing farmhouse and the guarantee that the "Blauzone" would not restrict the cultivation of agricultural land.

In total, 96 written appeals were made during the review and consultation process. The inputs included various, often overlapping aspects, which can be assigned to the following thematic priorities (Kopf 2016): agriculture (46%), property (21%), business development (9%), water management (7%) and other aspects (17%). Based on these submissions, further information and discussions occurred. In the end, some adjustments to the original plan were made, and approximately 2% (112 hectares) of the proposed area was further excluded, taking into account conflicting user interests. The state authorities regarded these minor changes as very justifiable (Interview 1).

Spatial Planning and Nature-Based Solutions in Flood Risk Management

The implementation process of the Blauzone Rheintal and the associated infringements of private property rights and future land-use opportunities illustrate the "growing importance of land resources in flood risk management" (Seher and Löschner 2017). By designating large-scale flood runoff and flood retention areas, the spatial plan regulates which land uses are compatible with the overall aim of providing more space for the Rhine and mitigating flood damage potential. In itself, regional planning instruments, such as the example discussed in this contribution, do not qualify as NBS, which are understood as "actions which are inspired by, supported by or copied from nature" (EC 2015). However, by securing the necessary land resources for the implementation of NBS, such as using wetlands to create emergency flood capacities, reconnecting rivers with floodplains or relocating dikes to make more space for flood storage, regulatory planning instruments provide important leverages for mitigating the impacts of floods and other natural hazards (EEA 2015).

Like NBS in general, the Blauzone Rheintal results in multiple co-benefits (EC 2015). Apart from reducing the risk of river flooding in the Rhine Valley, the spatial plan provides the following core benefits for the region:

- Preventing urban sprawl: By obliging municipalities to amend their local land use plans and (re)zone areas located within the "Blauzone", the planning instrument limits the encroachment of settlements into open space areas. Given the strong socio-economic growth in this part of the Rhine Valley, this supports planning strategies aimed at inner urban development.

- Securing agricultural production: The Blauzone Rheintal prevents settlement pressures on existing agricultural areas and secures productive soils along the Rhine for farming and agriculture. Although the "Blauzone" limits opportunities to expand agricultural buildings and financial profits from possible rezoning, the instrument only marginally interferes with existing land use rights and thus preserves future options for agricultural production.
- Preserving multifunctional open space: The "Blauzone" overwhelmingly coincides with the "Grünzone". It therefore, reinforces the legitimacy of an existing planning instrument and implicitly supports its fundamental goals, namely, the preservation of landscapes, biodiversity and local recreation areas.

The Blauzone Rheintal creates these multiple co-benefits in the Rhine Valley by assigning (long-term) land use and property rights (Needham and Hartmann 2016). Like other regulatory planning instruments, the spatial plan at first sight does not seem compatible with the "need for flexibility emanating from changing flood risk" (Tempels and Hartmann 2014) due to the "dynamic behaviour of floodplains as human-water systems" (Di Baldassarre et al. 2013). However, the "Blauzone" is explicitly designed as a rather flexible instrument in order to "adapt to changing future conditions" such as climate-induced changes in the flood hydrograph or to provide room for land development as a consequence of reduced flood hazards (Government of Vorarlberg 2013). By securing large-scale areas for future flood control measures (such as the relocation of the Rhine outlet into the Bodensee), the Blauzone Rheintal preserves space for manoeuvre to meet the long-term commitments associated with many measures in FRM (Interview 3). This forward-looking approach is best illustrated by the words of a leading architect of the Blauzone Rheintal: "Now we have secured these retention areas at least for the next generation. And if we had not done so, we would have seen buildings erected in these areas and emergency relief corridors would have been gone" (Interview 2).

In this regard, the Blauzone Rheintal is an example of a planning instrument that keeps options open for the future and prevents lock-in situations, which arise from floodplain development (Restemeyer et al. 2016). By preserving the flexibility of the system, the spatial plan also constitutes a reversible strategy in floodplain development (Hallegatte 2009): once flood protection measures (such as controlled retention basins) are implemented, policy makers may opt to reduce the extent of the "Blauzone" to again provide space for settlement development (Interview 2, Interview 3).

Despite the benefits of the planning instrument for adaptive, future-oriented FRM, the Blauzone Rheintal remains an isolated example of a regional approach to flood-related planning in Austria. As other countries have a more favourable institutional context for the implementation of risk reduction and FRM strategies in regional plans (Böhm et al. 2004), the Blauzone Rheintal is a planning instrument that is principally suitable for replication and upscaling in other flood-prone regions.

There are, however, some noteworthy contextual conditions of the Blauzone Rheintal, which should be taken into consideration in the case of adoption in other areas. First, the Blauzone Rheintal was developed on the basis of an existing, truly visionary spatial plan ("Grünzone"), which secured open areas and agricultural land as early as the late 1970s. It is highly doubtful whether the "Blauzone" could have been realized without the "Grünzone". Second, the spatial plan is part of large, transnational flood protection program (Rhesi). Its aims to provide emergency retention areas and secure areas for future flood control measures are facilitated by its being embedded in regional FRM. Third, due to the concentration of risk elements and the strong socio-economic growth in the Rhine Valley, settlement pressures necessitated a coordinated response at a larger scale to prevent further increases in damage potential. Finally, the planning instrument is the result of intensive cross-sectoral integration efforts by the state government departments of spatial planning and water management and a participatory process engaging various stakeholders, particularly the 22 municipalities and private landowners.

Acknowledgements We thank the Austrian Climate and Energy Fund for financing the projects Flood-Adapt (grant number: KR14AC7K11809) and RegioFlood (grant number: KR15AC8K12549) that led to the present book chapter.

Open access of this chapter is funded by COST Action No. CA16209 Natural flood retention on private land, LAND4FLOOD (www.land4flood.eu), supported by COST (European Cooperation in Science and Technology).

References

BfG (ed) (2015) Rheinatlas 2015. German Federal Institute of Hydrology (Bundesanstalt für Gewässerkunde—BfG), Koblenz

BMLFUW (ed) (2009) FloodRisk II. Vertiefung und Vernetzung zukunftsweisender Umsetzungsstrategien zum integrierten Hochwassermanagement. Synthesebericht. Bundesministerium für Land- und Forstwirtschaft, Umwelt und Wasserwirtschaft, Wien

BMLFUW (ed) (2015) Hochwasserrisikomanagementplan 2015. Risikogebiet: Alpenrhein 8001. Bundesministerium für Land- und Forstwirtschaft, Umwelt und Wasserwirtschaft, Wien

Böhm HR, Haupter B, Heiland P, Dapp K (2004) Implementation of flood risk management measures into spatial plans and policies. River Res Appl 20:255–267. https://doi.org/10.1002/rra.776

Di Baldassarre G, Kooy M, Kemerink JS, Brandimarte L (2013) Towards understanding the dynamic behaviour of floodplains as human-water systems. Hydrol Earth Syst Sci 17:3235–3244. https://doi.org/10.5194/hess-17-3235-2013

EC (ed) (2015) Towards an EU Research and Innovation policy agenda for Nature-Based Solutions & Re-Naturing Cities. Final report of the Horizon 2020 Expert Group on "Nature-Based Solutions and Re-Naturing Cities." European Commission, Brussels

EEA (ed) (2015) Exploring nature-based solutions. The role of green infrastructure in mitigating the impacts of weather- and climate change-related natural hazards. EEA technical report no 12/2015, European Environment Agency, Luxemburg

Frei R, Kopf M (2011) Blaue Zonen in Vorarlberg. Vorum—Forum Für Raumplan Reg Vorarlb 5–6

Fuchs S, Keiler M, Zischg A (2015) A spatiotemporal multi-hazard exposure assessment based on property data. Nat Hazards Earth Syst Sci 15:2127–2142. https://doi.org/10.5194/nhess-15-2127-2015

Government of Vorarlberg (2013) Blauzone Rheintal. Verordnung der Vorarlberger Landesregierung über die Festlegung von überörtlichen Freiflächen zum Schutz vor Hochwasser im Rheintal, Bregenz

Habersack H, Bürgel J, Kanonier A (2009) FloodRisk II: further steps for future implementation strategies towards an integrated flood risk management. English short report, Bundesministerium für Land- und Forstwirtschaft, Umwelt und Wasserwirtschaft, Wien

Hallegatte S (2009) Strategies to adapt to an uncertain climate change. Glob Environ Change 19:240–247. https://doi.org/10.1016/j.gloenvcha.2008.12.003

Hartmann T (2011) Contesting land policies for space for rivers—rational, viable, and clumsy floodplain management. J Flood Risk Manag 4:165–175. https://doi.org/10.1111/j.1753-318X.2011.01101.x

IRKA, IRR (eds) (2005) Entwicklungskonzept Alpenrhein—Kurzbericht. Internationale Regierungskommission Alpenrhein und Internationale Rheinregulierung

IRR (ed) (2017) Rhesi—Rhein, Erholung und Sicherheit. Internationale Rheinregulierug

Kidd S (2007) Towards a framework of integration in spatial planning: an exploration from a health perspective. Plan Theory Pract 8:161–181. https://doi.org/10.1080/14649350701324367

Klijn F, Samuels P, Os AV (2008) Towards flood risk management in the EU: state of affairs with examples from various European countries. Int J River Basin Manag 6:307–321. https://doi.org/10.1080/15715124.2008.9635358

Kopf M (2016) Strategie im Umgang mit Hochwasser: Blauzone Rheintal. Raumdialog Mag für Raumplan Reg Niederösterr 3(2016):20–21

Marik S (2005) Raumordnung in der Kommunalpolitik. Akteure—Entscheidungen—Umsetzung. Dissertation, Universität Wien, Fakultät Geowissenschaften, Geographie und Astronomie

Needham B, Hartmann T (2016) Planning by law and property rights reconsidered. Routledge, London

Petraschek A, Kienholz H (2003) Hazard assessment and mapping of mountain risks—example of Switzerland. In: Rickenmann D, Chen C (eds) Debris flow hazard mitigation. Millpress, Rotterdam, pp 25–38

Ran J, Nedovic-Budic Z (2016) Integrating spatial planning and flood risk management: a new conceptual framework for the spatially integrated policy infrastructure. Comput Environ Urban Syst 57:68–79. https://doi.org/10.1016/j.compenvurbsys.2016.01.008

Restemeyer B, van den Brink M, Woltjer J (2016) Between adaptability and the urge to control: making long-term water policies in the Netherlands. J Environ Plan Manag 60:920–940. https://doi.org/10.1080/09640568.2016.1189403

Samuels P, Klijn F, Dijkman J (2006) An analysis of the current practice of policies on river flood risk management in different countries. Irrig Drain 55:141–150. https://doi.org/10.1002/ird.257

Seher W, Löschner L (2017) Anticipatory flood risk management—challenges for land policy. In: Hepperle E, Dixon-Gough R, Mansberger R et al (eds) Land ownership and land use development. The integration of past, present, and future in spatial planning and land management policies. vdf Hochschulverlag AG, Zurich

State Government of Vorarlberg (2015) 10 Jahre danach. Jahrhundert-Hochwasser 2005. Maßnahmen und Strategie, Bregenz

State Government of Vorarlberg (2016a) Regionale Bevölkerungsprognose 2015 bis 2050, Bregenz

State Government of Vorarlberg (2016b) Integraler Hochwasserschutz. Risiken erkennen, vermindern, akzeptieren, Bregenz

Stead D, Meijers E (2009) Spatial planning and policy integration: concepts, facilitators and inhibitors. Plan Theory Pract 10:317–332. https://doi.org/10.1080/14649350903229752

Tempels B, Hartmann T (2014) A co-evolving frontier between land and water: dilemmas of flexibility versus robustness in flood risk management. Water Int 39:872–883. https://doi.org/10.1080/02508060.2014.958797

Warner JF, van Buuren A, Edelenbos J (2012) Making space for the river. IWA Publishing, London/New York

WKO (2017) Vorarlberger Regionen—Arbeitsplätze nach Wirtschaftssektoren 2012. Wirtschaftskammer Vorarlberg, Bregenz

Dr. Lukas Löschner has an academic background in Political Science and Landscape Planning and holds a Ph.D. in Spatial Planning. His main field of research is natural hazard risk management, which he explores in a combination of political science and planning approaches. He currently conducts research on the following topics: spatial adaptation to flooding, policy coordination in FRM, flood risk governance, and flood storage compensation.

Ass. Prof. Dr. Walter Seher holds a masters degree in Civil Engineering and Water Management and a Ph.D. from BOKU University. His key qualifications in research and teaching include spatial planning in natural hazard risk management and climate change adaptation, land rearrangement and land policy. Current research activities are focused on flood risk governance, spatial adaptation to flash floods and land policy in natural hazard risk management.

Dr. Ralf Nordbeck is a post-doc researcher at the University of Natural Resources and Life Sciences, Vienna (BOKU), and holds a Ph.D. in Political Science. Since 2005 he has worked and lectured at BOKU in the fields of environmental and natural resource policies. His current research interests are the comparative analysis of environmental policies, new modes of environmental governance, flood risk governance, and intersectoral and multi-level coordination issues.

Dipl.-Ing. Manfred Kopf studied Landscape Ecology and Landscape Design at BOKU University and expanded his spatial planning knowledge in postgraduate studies at ETH Zurich. As head of the division for Regional Spatial Planning in the State Government of Vorarlberg, he is primarily involved with the implementation of spatial planning policies for settlement development and the conservation of open space.

Chapter 16
Commentary: The Blauzone Rheintal from a Hydrological Perspective—A Transboundary Flood-Mitigation Solution in a Large Gravel-Bed River Basin

Nejc Bezak, Mojca Šraj and Matjaž Mikoš

Hydrological Perspective: Reflection and Open Questions

The case of the Alpine Rhine (Alpenrhein in German) upstream of the Lake of Constance (Bodensee in German) is a good example for a large alpine gravel-bed river that is shared by several countries and for high population density from the Alps perspective. Bringing into focus also common transboundary issues in water management of large basins shared by neighbouring countries and presenting open dilemmas due to different water management and spatial planning policies and procedures in these countries, the presented work (i.e., named "Blauzone Rheintal" in the book chapter written by Löschner et al. this volume) can be regarded as an advanced case study. Why? Firstly, the Alpine Rhine was heavily regulated in the past (Meyer-Peter and Lichtenhahn 1963; Lichtenhahn 1972). Secondly, the Alpine Rhine is a part of the large European Rhine River Basin, where in the field of water management the international cooperation is lead by the International Commission for the Protection of the Rhine (ICPR 2018), and between Switzerland and Austria, since 1892, the Internationale Rheinregulierung (IRR 2018a). Over the decades, these different international bodies have prepared very good hydrologic, hydraulic, and other flood-hazard related data, models and concepts such as the 2005 Alpine Rhine concept (IRKA and IRR 2005) and the Rhesi flood protection project (IRR 2018b) that presented a very good basis for the implementation of the Blauzone. As noted by Löschner et al. (this

N. Bezak (✉) · M. Šraj · M. Mikoš
Faculty of Civil and Geodetic Engineering, University of Ljubljana, Ljubljana, Slovenia
e-mail: nejc.bezak@fgg.uni-lj.si

M. Šraj
e-mail: mojca.sraj@fgg.uni-lj.si

M. Mikoš
e-mail: matjaz.mikos@fgg.uni-lj.si

T. Hartmann et al. (eds.), *Nature-Based Flood Risk Management on Private Land*,
https://doi.org/10.1007/978-3-030-23842-1_16

volume), the Blauezone is part of the Rhesi flood protection project. Rhesi is flood protection project that can be regarded as the first major phase in the implementation of the Alpine Rhine development concept (IRR 2018b). Moreover, the Blauezone (blue zone) was implemented based on the spatial plan named Grünzone (green zone) that was developed almost 50 years ago and secured areas for agricultural purpose in the Alpine Rhine valley (Löschner et al. this volume). These efforts have sought regional flood risk reduction in the Alpine Rhine valley by implementing different structural and non-structural measures (Stalzer 2007, 2008). From a hydrological perspective, for the development of models and concepts very good input data (e.g., appropriate density of rainfall gauging stations, high-frequency measurements) is needed, and state-of-the-art methods should be used with the consideration of uncertainty in the model results. Morevoer, possible climate change or variability impacts should also be taken into consideration in hydrological investigation.

The Alpine climate zone is an area with very high soil loss rates due to water erosion because of the combined effect of topography (i.e., steep slopes) and rainfall erosivity (e.g., summer thunderstorms) (Panagos et al. 2015). Moreover, in areas located above the tree line, the vegetation cover is sparse, which, in combination with other erosion types (e.g., ice erosion), additionally increases erosion rates. Thus, high soil erosion rates and intense and complex sediment transport processes can be expected in Alpine rivers, which means that not only floods but also a combination of sediment transport and corresponding morphological processes can endanger people leaving in the Alpine valleys (Fäh et al. 2012). The Alpine Rhine was known as the largest torrent in Europe, and, after several extreme events that occurred between 1801 and 1900, the IRR was established (Fäh et al. 2012). First measures that were taken did not take an interdisciplinary approach and were not able to solve the flooding issue in the valley. They also influenced bed-load balance and groundwater level (Zarn 2008; Fäh et al. 2012). However, newer interdisciplinary approaches also considered bed load management in combination with other aspects, such as ecology and development of recreational activities (IRKA and IRR 2005). In the past, due to gravel mining, at some sections, the river bed had decreased by almost 5 m between 1941 and 1995 (Zarn 2008). The eroded material was deposited in the Lake of Constance and led to delta growth (Zarn 2008). At the moment, the river bed at the international section of the Alpine Rhine is near a stable state of equilibrium (Zarn et al. 1995; Fäh et al. 2012), which means that a potential change in river cross-section can lead to aggradation or degradation of the river bed, which can also have an impact on flood safety. Thus, other approaches, such as spatial planning, are needed in order to ensure flood safety in this area. The importance of erosion and sedimentation processes in river basins was also underlined by the International Sediment Initiative (ISI 2018) launched by the United Nations Educational, Scientific and Cultural Organization (UNESCO) International Hydrological Programme (IHP) in 2004. Therefore, it is important to stress that, from a hydrological point of view, in such a large alluvial river as the Alpine Rhine, flood risk assessment and countermeasures must also take river sediments and estimation of sediment budget of a river into account (Stalzer 2007, 2008).

Changing flood peaks by controlled overflow of dikes and routing flood waters into designated flood areas (agricultural land, riverine forests, wetlands, …) may change the sediment regime in the main watercourse and, in flooded areas, may cause local soil erosion problems and/or accumulation of fine sediments by settling of suspended sediments in flood waters. Furthermore, from an engineering perspective, the design of water routing to the designated flood areas (i.e., hydro-technical structures used for water diversion) is a complicated task since the water velocities during the flood events are high (cca. between 1 and 5 m/s) when these areas are actively used. In combination with high bed and suspended load transport during flood events, this makes it difficult to predict water levels (i.e., large uncertainty) and consequently predict water movement on the designated flood areas (i.e., issue of effective use of flood volume). Moreover, this kind of structures also cannot be tested before the flood event (i.e., operational testing). The issue of flood damages on the designated flood areas determined through spatial planning procedures must be regulated in advance. Therefore, how spatial planning of flood risk reduction measures are conveyed to stakeholders is of paramount importance as well as how flood risk dialogue is conceptualised and put into practice. The Blauzone Rheintal is a good example, how, as a part of advanced water management concepts for flood risk reduction in the Alpine environment, spatial planning should be applied as a tool to bring such concepts to life. From this perspective, it is a case study that discusses the structure and responsibilities on a state and regional level in spatial planning procedures that are not the same across multiple countries. How adaptive water management concepts are turned into practice is also to a large extent dependent on how land use policy is regulated by spatial planning acts and who is responsible for planning—the state or local communities (municipalities). In the field of water management, the European Water Directive and the Flood Directive ask for an integrated basin-level approach in flood hazard assessment and flood risk reduction policy (for the Alpine Rhine see Stalzer 2007, 2008). The way to fulfil planned flood risk reduction measures at different scales is to have clear spatial planning instruments (procedures, regional and local plans) that are capable of effectively implementing water management measures.

Another process that is important from a hydrological perspective and was also investigated and considered in the Alpine Rhine case study is groundwater recharge (e.g., Zarn 2008). Changes in the Alpine Rhine River bed in the previous century also influenced the groundwater level that decreased in the past (Zarn 2008). Since groundwater is an important source of drinking water for almost half a million inhabitants in the Alpine Rhine valley (Zarn 2008), the proposed and implemented flood protection measures that stabilised the river bed consequently have a positive impact on groundwater storage. Moreover, the use of designated flood areas (i.e., Blauzone) for flood protection also has a positive impact on the groundwater storage since the flood water is not routed directly to Lake Constance but it is kept in the Alpine Rhine catchment for a longer time period (i.e., infiltration to the soil and groundwater recharging).

Transferability Across Scales, Places and Disciplines and Final Remarks

The Sava River (part of the Danube River Basin) is another case study where sediment management is considered as part of the FRM plan. In the Sava River, the efforts for integrated water management in this basin are led by the International Sava River Basin Commission (ISRBC 2018). The ISRBC also prepared the Sava River Basin Management Plan (Sava RBMP 2018) and the FRM Plan in the Sava River Basin (Sava FRMP 2018). However, the implementation of spatial planning measures is up to countries and cannot be fully regulated on the international scale.

Moreover, the described Alpine Rhine case study (Zarn 2008; Löschner et al. this volume) is aligned with the integrated water management policy and is especially pertinent to the flood risk reduction policy that calls for "more room for water"—which has been more and more widely accepted; the Dutch project "Room for the River" may serve as an example here. In 2006, the Dutch Cabinet proposed the Spatial Planning Key Decision (SPKD)—a design plan for more highly innovated structures and the modification of existing structures in the immediate floodplains of four major Dutch rivers—the project was executed from 2006 to 2015 (Room for the River 2018).

It is clear that the Rhesi flood protection project (part of which is a Blauzone regional planning instrument) can be regarded as an interdisciplinary approach that considered numerous aspects of flood protection (e.g., planning of recreation activities in the area, sediment transport, groundwater storage, etc.). Moreover, intense transnational cooperation was also needed to implement Rhesi and Blauzone. Thus, the presented case study could be transferred to different areas and to smaller and larger catchments than the Alpine Rhine catchment. In a different situation, the area along the river could be more densely populated, which would enable implementation of measures such as Blauzone regional planning that was based on the older measure that secured agricultural land along the river (Löschner et al. this volume). Moreover, this kind of approach is probably not very suitable for smaller Alpine rivers with significant torrential characteristics, because this kind of river is usually located in narrow valleys without much free space for dedicating land to water. Furthermore, torrents have a very fast response; this complicates the design of designated flood areas. Finally, the presented approach can be regarded as state-of-the-art and can serve as a role model for other alpine catchments.

Acknowledgements The results of the study are part of the research Programme P2-0180: "Water Science and Technology, and Geotechnical Engineering: Tools and Methods for Process Analyses and Simulations, and Development of Technologies" that is financed by the Slovenian Research Agency (ARRS).

Open access of this chapter is funded by COST Action No. CA16209 Natural flood retention on private land, LAND4FLOOD (www.land4flood.eu), supported by COST (European Cooperation in Science and Technology).

References

Fäh R, Weiss M, Hengl M, Dietsche D, Boes R (2012) Alpine Rhine river—sustainable flood protection measures between the levees. In: 12th congress interpraevent 2012 proceedings, Grenoble, France, pp 23–26, April 2012

ICPR (2018) International Commission for the Protection of the Rhine. DIALOG. www.iksr.org. Accessed 8 Aug 2018

IRKA, IRR (2005) Entwicklungskonzept Alpenrhein – Kurzbericht. DIALOG. http://www.rhy-faescht.org/fileadmin/user_upload/2015/rhyfaescht-data/b_documents/IRKA_EKA_Alpenrhein_2005_kurz.pdf. Accessed 8 Aug 2018

ISI (2018) International Sediment Initiative. DIALOG. http://his.irtces.org/isi/. Accessed 8 Aug 2018

ISRBC (2018) International Sava River Basin Commission. DIALOG. www.savacommsion.org. Accessed 8 Aug 2018

IRR (2018a) Internationale Rheinregulierung. DIALOG. www.rheinregulierung.org/. Accessed 8 Aug 2018

IRR (2018b) Rhesi - Rhein, Erholung und Sicherheit. Internationale Rheinregulierug. DIALOG. http://www.rhesi.org/. Accessed 8 Aug 2018

Lichtenhahn C (1972) Flussbauliche Probleme am Rhein zwischen Reichenau und dem Bodensee im Wandel der Zeit. Wasser- und Energiewirthschaft 64(10–11):341–353

Meyer-Peter E, Lichtenhahn C (1963) Altes und Neues über den Flussbau. Eidg. Departement des Innern, Veröffentlichungen des Eidg. Amtes für Strassen- und Flussbau, Bern

Panagos P, Borrelli P, Poesen J, Ballabio C, Lugato E, Meusburger K, Montanarella L, Alewell C (2015) The new assessment of soil loss by water erosion in Europe. Environ Sci Policy 54:438–447

Room for the River (2018) Room for the River for a safer and more attractive river landscape. DIALOG. https://www.ruimtevoorderivier.nl/english/. Accessed 9 Aug 2018

Sava FRMP (2018) Flood risk management plan in the Sava River Basin. DIALOG. http://www.savacommission.org/sfrmp/en/. Accessed 8 Aug 2018

Sava RBMP (2018) Sava River Basin Management Plan. DIALOG. http://www.savacommission.org/srbmp/en/. Accessed 8 Aug 2018

Stalzer W (2007) Alpenrhein 2100 – Vom Gestern zum Morgen im Alpenrheintal. Oesterr Wasser Abfallwirtsch 59(7–8):23–28

Stalzer W (2008) Der Alpenrhein – Versuch einer nachhaltigen Entwicklung für den gröβten alpinen Wildfluss. Oesterr Wasser Abfallwirtsch 60(5–6):73–79

Zarn B (2008) Development concept river Alpine Rhine. In: 11th congress interpraevent 2008 Proceedings, Dornbirn, Austria, pp 26–30, May 2008

Zarn B, Oplatka M, Pellandini S, Mikoš M, Hunziker R, Jäggi M (1995) Geschiebehaushalt Alpenrhein. Mitteilungen der VAW 139, ETH Zürich

Nejc Bezak Ph.D., is Assistant Professor at the Faculty of Civil and Geodetic Engineering, University of Ljubljana. He is member of the UNESCO Chair on Water-Related Disaster Risk Reduction. His main research interests are various hydrology sub-fields such as statistical hydrology (e.g., multivariate analyses, trend detection, design discharge determination, flood risk assessment), hydrological and hydraulic modelling, data-analyses, climate change investigation, field measurements.

Mojca Šraj Ph.D., is Associate Professor of Hydrology at the Faculty of Civil and Geodetic Engineering, University of Ljubljana and a member of the UNESCO Chair on Water-Related Disaster Risk Reduction. Her main research interests include applied hydrology, rainfall-runoff modeling, extreme events analyses, flood risk assessment, forest hydrology, time series analyses and climate change. She has published more than 90 scientific papers in the area of hydrology and water resources.

Matjaž Mikoš dr. sc. techn. ETH Zurich, is Professor of Hydrology and Hydraulic Engineering, currently Dean of the Faculty of Civil and Geodetic Engineering, University of Ljubljana, and Head of the UNESCO Chair on Water-Related Disaster Risk Reduction. He is Member of the INTERPRAEVENT Scientific-Technical Board; he served two terms as its head. He is expert in applied hydrology, sediment management, river engineering, torrent control, and flood and landslide risk management.

Chapter 17
Commentary: The Blauzone Rheintal Approach from a Natural Hazard Perspective—Challenges to Establish Effective Flood Defence Management Programs

Carla S. S. Ferreira and Zahra Kalantari

Flood Hazard and Nature-Based Solutions for Mitigation Measurements

Water-based natural and human-induced disasters have been increasingly affecting communities worldwide, with single events causing extraordinary adverse social, economic and environmental consequences. Floods are among the most frequent hazards and led to an average annual loss in the EU of EUR 4.9 billion for the period 2000–2012, and may reach EUR 23.5 billion by 2050 (Jongman et al. 2014). Increasing vulnerability of population and infrastructures to floods is mainly driven by (i) urbanization trends, since the expansion of paved surfaces affects the hydrological cycle, particularly due to reduced infiltration and faster peak flows (e.g., Ferreira et al. 2018); and (ii) climate change, which will alter, among others, precipitation patterns, resulting in more intense and frequent storms. These trends have received growing attention on a global scale (Abdulkareem and Elkadi 2018), particularly because conventional flood control structures, based on technical and engineering-dominated approaches, are increasingly questioned among academics, decision makers and communities. Over the last years, a new approach for FRM has been rising, based on principles of resilience found in the natural world (Abdulkareem and Elkadi 2018). Mitigation and non-structural solutions tend to be potentially more efficient and sustainable to water-related problems, re-directing the focus away from struc-

C. S. S. Ferreira (✉)
Research Centre for Natural Resources, Environment and Society (CERNAS), Polytechnic Institute, Coimbra College of Agriculture, Coimbra, Portugal
e-mail: carla.ssf@gmail.com

Z. Kalantari
Department of Physical Geography, Bolin Centre for Climate Research, Stockholm University, Stockholm, Sweden
e-mail: zahra.kalantari@natgeo.su.se

T. Hartmann et al. (eds.), *Nature-Based Flood Risk Management on Private Land*,
https://doi.org/10.1007/978-3-030-23842-1_17

turally protective measures towards spatial planning and NBS (e.g., Kalantari et al. 2018).

The Blauzone Rheintal approach, in Austria, which involves establishing "blue zones" in local land-use plans, is an example of the "growing importance of land resources in flood risk management" (Seher and Löschner 2017). The Blauzone Rheintal system creates multiple co-benefits by securing the necessary land resources for implementation of NBS, such as using wetlands to create emergency flood capacity, reconnecting rivers with floodplains or relocating dikes to make more space for flood storage. It also contains regulatory planning instruments that provide important leverage for mitigating the impacts of floods and other natural hazards.

The potential of NBS is beginning to reframe the discussions and policy responses to implement mitigation measures against extreme water-based events. The term NBS is used here to describe soft engineering approaches that are aimed at increasing the resilience of territories and societies affected by meteorological events (Potschin et al. 2016). The NBS concept builds on and recognizes the importance of nature and highlight the requirement for a systemic and holistic approach to environmental change based on an understanding of the structure and functioning of ecosystems and the social ownership and institutional context within which they are situated. Nature-based solutions can be flexible and multi-beneficial for ecosystems, providing an alternative and/or complement to conventional engineering flood defence strategy. The EU has been devoted to support an approach involving a combination of defensive actions and adaptive management of natural resources (EU 2015).

Background Information to Establish Effective Measures to Prevent Flood Hazard

Strategies for adapting to increasing flood risks and climate change should focus on prevention, protection and preparedness, as highlighted by the Floods Directive (2007/60/EC), and must be defined by a range of environmental and social factors (Loos and Rogers 2016).

A flood defence strategy should build upon flood hazard maps and flood risk maps, considering several flood scenarios including different possibilities such as return periods and land-uses, showing the potential adverse consequences of floods. The use of tools integrating Geographic Information Systems with hydrological modelling are very useful to prepare these maps, but the uncertainty of the results should not be ignored. Models require data calibration and validation based on long records of hydrological measurements, as well as a description of past floods. These data are not always available, raising uncertainties in flood scenarios that are not easy to quantify (Ceppi et al. 2010).

Flood risk management plans should focus on the potential to retain water in the landscape (e.g., Ferreira et al. 2015) and consider "more space for rivers". The use of "blue" and "green" spaces to tackle runoff and other management issues, such as attenuation of peak runoff and water quality, is a less expensive and more long-lasting solution than "gray" infrastructures (e.g., Lafortezza et al. 2018). However, quantification of the effectiveness of existing large-scale NBS like Blauzone Rheintal, their operationalization and replicability in different local situations needs to be performed in a way that allows NBS to be both widely accepted and incorporated into policy development and practical implementation. Therefore, developing, implementing and upscaling NBS like Blauzone Rheintal requires gathering appropriate quantitative and qualitative information and utilizing this information to answer three key questions:

- Is the proposed NBS effective (economically, physically and socially) in addressing the problem (e.g., reducing hydro-meteorological hazards and climate change adaptation at watershed/landscape scale)?
- What are the most successful regulatory frameworks and inclusive management approaches for implementing and maintaining an effective NBS in a sustainable manner?
- How do the effectiveness, operational requirements and sustainability of the proposed NBS change with local conditions?

The range of evidence required to assess the effectiveness of NBS includes the bio-physical and economic aspects, social and behavioural data on levels of understanding, acceptance, implementation and sustainable management, and information about cost-effectiveness. With strong positive evidence, acceptance for NBS as risk-reduction strategies will become feasible, and they can be adopted by local communities with several advantages for ecosystem services (recreation, biodiversity, education, etc.).

Social and Environmental Aspects Contributing to Flood Management

Other countries may learn from cases like Blauzone Rheintal, which developed a regulatory instrument for FRM based on integration of different policy domains, particularly linked with water management and spatial planning, and involving public and private sector activities. This instrument incorporates sustainability aspects to prevent and reduce damages to human health, environment, economic activities and cultural heritage. Nature-based solutions often involve working with local communities, landowners, land managers and risk management officers, in order to achieve the threshold of intervention that is effective in reducing the risks from water-based hazards (SNH 2010).

In Sweden, the Kristianstad municipality in the Helge River basin is one of the front-runners regarding hydro-meteorological risk management. Situated in the middle of a wetland area, Kristianstad city has both struggled and developed together with the Helge River. Flooding is a problem in both rural and urban areas. In fact, Kristianstad is the most flood-prone town in Sweden, with parts of the town almost

2.5 m below sea level, and historical land-use actions such as embankments, lowering of lakes, straightening of flows and dredging, have reduced the wetland area and the landscape's capacity to retain water. Rural parts of the basin (e.g., agricultural land and wetlands) thus face a recurring flood risk due to annual fluctuations in river water level. Embankments protect some more low-lying agricultural areas, but the land is vulnerable to severe flooding if they fail. The city itself is vulnerable to flows coming from upstream in the river basin, and when these flows are dammed back from the Baltic Sea by low land elevation, the drainage speed is reduced. With sea-level rise this damming effect will be greater, posing a high flood risk to Kristianstad. The city has long been working with strategies to prevent flooding, mainly involving structural measures (e.g., embankments). But during recent decades, the regime in Kristianstad has changed to a living-with-water approach, and the municipality is now looking into other possible future strategies, rather than business-as-usual (Johannessen and Hahn 2013). The Blauzone Rheintal regional plan for designating large-scale areas for flood retention and flood runoff can be relevant and practicable in the Kristianstad municipality. Such a regional plan in the area could include wetlands, sandy arable soils and meadows, developed in ways that benefit both the natural environment and human beings. For example, flooded grasslands could act as buffer zones for flood and nutrient-leaching prevention, and also provide new habitats for bird conservation. This new strategy will allow flooding of certain areas, but land-use is adapted to minimize its impact. The strategy also means synergies with nature and landscape development, having the "river as a partner" and benefiting from ecosystem services.

The importance of stakeholder engagement and participation in FRM has been recognized in several countries, such as the Netherlands and the United States (Loos and Rogers 2016). The involvement of local stakeholders in decision-making provides (i) improvement of the quality of decision making by using their information and solutions, since they are more acutely aware of specific vulnerabilities to flood and climate change, as well as local economic, environmental and social conditions (e.g., Loos and Rogers, 2016); and (ii) democratic legitimacy and trust, in order to enhance acceptance of decision outcomes (Edelenbos et al. 2017). Instead, public participation processes, as considered by other countries, are rather a simple process of hearing public comments and have left many stakeholders dissatisfied (Loos and Rogers 2016).

Lessons Learnt from Blauzone Rheintal

After decades of neglect, the importance of protecting and improving ecosystems for reducing disaster risk has started to receive attention in recent years (Gupta and Nair 2012). Considering the trends and changes in hydro-meteorological events, the approach to deal with natural hazards requires a change of paradigm, shifting from an approach relying exclusively on defensive action against natural hazards to an approach combining defensive action with adaptive management of natural resources (Cohen-Shacham et al. 2016). The ambition of the Blauzone Rheintal system is to facilitate this change in paradigm through operationalizing NBS, by designating large-scale areas for FRM, through intersectoral coordination and engagement of local stakeholders. There is international acknowledgement that efforts to reduce disaster risks must be systematically integrated into policies, plans and programs for sustainable development, and supported through bilateral, regional and international cooperation, including partnerships, particularly in the case of cross-border watersheds (ISDR 2005). A FRM program such as the Blauzone Rheintal offers new opportunities and brings added value, encompassing ideas and providing inclusion of local knowledge and lessons from the past (Eggermont et al. 2015). It is important to use all relevant sources of expertise in developing and implementing NBS like Blauzone Rheintal, and to identify the lessons learnt within and across the project (e.g., Nesshöver et al. 2017).

The NBS approach involves working with nature and, in essence, aims at increasing the natural capital of ecological systems, for example, reducing flood risk. To be successful, NBS must consider local social-ecological systems, so that local communities, landowners and land managers are engaged, in order to secure a threshold of interventions that can effectively reduce the risks from water-based disasters (SNH 2010). The concept underpinning NBS builds on, and complements, other concepts such as the ecosystem approach, ecosystem services, ecosystem-based adaptation/mitigation, disaster risk reduction, and green and blue infrastructure. All of these recognize the fundamental importance of working with nature and using a systemic, holistic approach to environmental change based on an in-depth understanding of the structure and functioning of ecosystems, and the social and institutional context within which they are situated.

The Blauzone Rheintal program was developed by distinct institutions which coordinate and support water management authorities, in cooperation with planners and decision makers. This approach provides effective long-term strategic plans for districts, cities or regions. A broader application of NBS into FRM requires integration among researchers, politicians and the economic sector, in order to provide evidence-base for NBS cost-effectiveness, co-benefits and up scaling benefits.

Acknowledgements Open access of this chapter is funded by COST Action No. CA16209 Natural flood retention on private land, LAND4FLOOD (www.land4flood.eu), supported by COST (European Cooperation in Science and Technology).

References

Abdulkareem M, Elkadi H (2018) From engineering to evolutionary, an overarching approach in identifying the resilience of urban design to flood. Int J Disaster Risk Reduct 28:176–190. https://doi.org/10.1016/j.ijdrr.2018.02.009

Ceppi A, Ravazzani G, Rabuffetti D, Mancini M (2010) Evaluating the uncertainty of hydrological model simulations coupled with meteorological forecasts at different spatial scales. Procedia Soc Behav Sci 2:7631–7632. https://doi.org/10.1016/j.sbspro.2010.05.152

Cohen-Shacham E, Walters G, Janzen C, Maginnis S (eds) (2016) Nature-based solutions to address global societal challenges. IUCN, Gland, Switzerland, xiii + 97 pp

Edelenbos J, Buuren AV, Roth E, Winnubst M (2017) Stakeholder initiatives in flood risk management: exploring the role and impact of bottom-up initiatives in three 'Room for the River' projects in the Netherlands J Environ Plann Manag 60(1):47–66. https://doi.org/ 10.1080/09640568.2016.1140025

Eggermont H, Balian E, Azevedo JMN, Beumer V, Brodin T, Claudet J (2015) Nature-based solutions: new influence for environmental management and research in Europe. GAIA—Ecol Perspect Sci Soc 24:243–248. http://dx.doi.org/10.14512/gaia.24.4.9

EU (2007) Directive 2007/60/EC of the Parliament and the Council of 23 October 2007 on the assessment and management of flood risks. Off J Eur Union L288/27–L288/34

EU (2015) Towards an EU Research and Innovation policy agenda for Nature-Based Solutions & Re-Naturing Cities. Final report of the Horizon 2020 Expert Group on 'Nature-Based Solutions and Re-Naturing Cities', Directorate-General for Research and Innovation, Climate Action, Environment, Resource Efficiency and Raw Materials EN

Ferreira CSS, Walsh RPD, Steenhuis TS, Shakesby RA, Nunes JPN, Coelho COA, Ferreira AJD (2015) Spatiotemporal variability of hydrologic soil properties and the implications for overland flow and land management in a peri-urban Mediterranean catchment. J Hydrol 525:249–263. https://doi.org/10.1016/j.jhydrol.2015.03.039

Ferreira CSS, Walsh RPD, Steenhuis TS, Ferreira AJD (2018) Effect of peri-urban development and lithology on streamflow in a Mediterranean catchment. Land Degrad Dev 29:1141–1153. https://doi.org/10.1002/ldr.2810

Gupta AK, Nair SS (2012) Ecosystem approach to disaster risk reduction. National Institute of Disaster Management, New Delhi, India, p 202

ISDR (2005) Hyogo framework for action 2005–2015: building the resilience of nations and communities to disasters extract from the final report of the World Conference on Disaster Reduction (A/CONF.206/6), Kobe, Hyogo, Japan, 18–22 January 2005

Johannessen Å, Hahn T (2013) Social learning towards a more adaptive paradigm? Reducing flood risk in Kristianstad municipality, Sweden. Glob Environ Change 23(1):372–381. https://doi.org/10.1016/j.gloenvcha.2012.07.009

Jongman B, Hochrainer-Stigler S, Feyen L, Aerts JCJH, Mechler R, Botzen WJW, Bouwer LM, Pflug G, Rojas R, Ward P (2014) Increasing stress on disaster-risk finance due to large floods. Nat Clim Change 4:264–268. https://doi.org/10.1038/NCLIMATE2124

Kalantari Z, Ferreira CSS, Keesstra S, Destouni G (2018) Nature-based solutions for flood-drought risk mitigation in vulnerable urbanizing parts of East-Africa. Curr Opin Environ Sci Health 5:73–78. https://doi.org/10.1016/j.coesh.2018.06.003

Lafortezza R, Chen J, van den Bosch CK, Randrup TB (2018) Nature-based solutions for resilient landscapes and cities. Environ Res 165:431–441. https://doi.org/10.1016/j.envres.2017.11.038

Loos JR, Rogers SH (2016) Understanding stakeholder preferences for flood adaptation alternatives with natural capital implications. Ecol Soc 21(3):32. https://doi.org/10.5751/ES-08680-210332

Nesshöver C, Assmuth T, Irvine KN, Rusch GM, Waylen KA, Delbaere B, Haase D, Jones-Walters L, Keune H, Kovacs E, Krauze K, Külvik M, Rey F, van Dijk J, Vistad OI, Wilkinson ME, Wittmer H (2017) The science, policy and practice of nature-based solutions: an interdisciplinary perspective. Sci Total Environ 579:1215–1227. https://doi.org/10.1016/j.scitotenv.2016.11.106

Potschin M, Kretsch C, Haines-young R, Furman E, Berry P, Baró F (2016) Nature-based solutions. OpenNESS 1604 ecosystem service reference book

Seher W, Löschner L (2017) Anticipatory flood risk management—challenges for land policy. In: Hepperle E, Dixon-Gough R, Mansberger R et al (eds) Land ownership and land use development. The integration of past, present, and future in spatial planning and land management policies. vdf Hochschulverlag AG, Zurich

SNH (2010) Flood management. Available via DIALOG. http://www.snh.gov.uk/land-and-sea/managing-freshwater/flooding. Accessed 08 March 2019

Carla Ferreira is a post-doctoral researcher in the Research Center of Natural Resources, Environment and Society (CERNAS). Her main field of research is land degradation, particularly driven by land use changes. She is an expert in assessing the impacts of urbanization on catchment hydrology and sediment fluxes. Her research focus also revolves around understanding the role of landscape and land-use planning, among other NBS, to mitigate hydro-climatic risks such as floods.

Zahra Kalantari is Associate Professor at Stockholm University and Research Area Co-Leader of the Bolin Centre for Climate Research. Her special field of research is applied science on climate change mitigation and management of the impact of land use changes on catchment hydrology. She is an expert on vulnerability assessment to water-related disasters, adaptive land-use planning for decision support, and valuation of NBS for sustainable rural and urban development.

Part VII
Dike Relocation in Germany

Chapter 18
Relocation of Dikes: Governance Challenges in the Biosphere Reserve "River Landscape Elbe-Brandenburg"

Barbara Warner and Christian Damm

As model regions for sustainable development, biosphere reserves have to protect large landscape units on the background of uncertain effects of climate change. They have to implement a suitable management for sustainable development of nature, and they have to include many stakeholders in their governance-processes. Long-lasting solutions for conservation, flood protection and socioeconomic approaches are required.

The focus of this article lies on a riparian landscape area, the biosphere reserve Elbe-Brandenburg River Landscape. Against the background of strained socioeconomic conditions and many requirements from stakeholders, the project "dike relocation near Lenzen (Brandenburg)" had to secure and improve the ecological conditions of floodplains, including forests, largely by turning private (agricultural) land back to floodplain forests and other floodplain-specific habitat types. Achieving acceptance for the goals of the project among the stakeholders and the residents and the need for suitable land use on a large scale were of great importance. The consequent involvement of different stakeholders was essential for a successful project combining ecological restoration and flood mitigation.

B. Warner (✉)
Academy for Spatial Research and Planning (ARL), Leibniz-Forum for Spatial Sciences, Academic Section Ecology and Landscape, Hannover, Germany
e-mail: warner@arl-net.de

C. Damm
Department of Wetland Ecology, Karlsruhe Institute of Technology (KIT)/Institute of Geography and Geoecology, Karlsruhe, Germany
e-mail: christian.damm@kit.edu

T. Hartmann et al. (eds.), *Nature-Based Flood Risk Management on Private Land*,
https://doi.org/10.1007/978-3-030-23842-1_18

Introduction

Flood retention and ecological demands have their specific relevance concerning climate change. Adaptive nature conservation management is affected by a lack of sufficient understanding of the functional relations in natural systems (and therefore in the conservation areas as well). It is affected also by current uncertainties concerning the impacts of climate change. For the implementation of large projects concerning flooding, linkages between different actors or stakeholders and their specific approaches are to be taken into account.

Climate Change and Land Protection

Climate change requires flexible solutions. The inherent uncertainty contradicts the demand for specificity and unambiguity in planning, and this affects strategies of spatial (landscape) planning. Planning instruments and strategies in Germany are not suited for the flexibility required for climate change adaptation. Planning with flood scenarios, for example—as the European Floods Directive requires (see Hartmann and Spit 2016)—has an impact on the use of instruments like Strategic Environmental Assessment. For some consequences of climate change, even scenarios are not suitable. Instead they require individual adaptation of strategies or instruments. This causes problems in areas or situations when different interests have to be weighed—as this is normally a standard for integrative managing in conservation areas. What does climate change mean for the protection grounds? As an effect of renaturation of meadows, for example, biodiversity is strengthened and fortifies ecosystems against decline with its variety of species ("insurance hypothesis", see Yachi and Loreau 1999).

The dilemma between the legal certainty of planning instruments and the requirement of flexibility asks for different ways to use (planning) instruments. Communication and "good governance" are the key for this challenge (Kreibich et al. 2011) and important to implementing climate adaptation.

Dike Relocation as a Challenge for Nature and Governance

But how to organize the process of adaptation to climate change in concrete fields of action? For conservation areas, recommendations for adaptation are not specific enough to support their management with its special characteristics and communicative challenges. Narrow compartmentalization of responsibilities may strengthen the capacity to act for administration, state actors or municipalities; on the other hand, it encourages sectoral breakdowns of challenges and leads to thinking in delimited areas.

Could an ecosystem-based approach lead to win–win-situations or does the approach stay within the limits of the specific professional responsibilities (Warner and Rannow 2016)? Conservation strategies for the regional or national level have to consider the specific management practice. A practice-oriented approach for strategies is required, which includes species conservation, policy, law and governance, land and water management and protection, research, knowledge and science, involvement of local stakeholders, citizens and external or local experts.

Some basic questions could be figured out concerning a process of adaptation of climate change (Warner and Rannow 2016, shown in Wilke and Rannow 2013): Where could suitable and specific information about potential effects of climate change be found? And how should different stakeholders be included?

The example of a dike relocation project in the biosphere reserve Elbe-Brandenburg River Landscape shows a successful management of demands concerning protection (nature and land), properties (land use strategies and means) and participation (governance). To enable flooding, dikes had to be remodeled on a large scale. This process affected mainly agricultural land. The main challenge was to convince farmers to sell their land, and to keep them as stakeholders in the whole project.

Fostering acceptance in a remote region of East Germany was one challenge and of major importance for the project. It was solved mainly by a close cooperation between the administration of the biosphere reserve and the local agricultural holding company as the sole tenant of most of the agricultural acreage in the project area. With state subsidies and a powerful NGO's support, it was possible to create and secure large flooded areas—based on "good governance" with all parties.

Main Points and Structure

The article identifies challenges for the planning of and for large protected areas. It uses the example of a planning project in northeastern Germany to show how flood protection can be implemented. We show how a combination of ecological restoration and flood mitigation has been successfully realized. The project "dike relocation near Lenzen (Brandenburg)" illustrates how land consolidation schemes and land users' participation could be used to obtain land availability. The planning phase started in the 2000s, and acceptance among the local public was a main barrier to implementing the dike relocation part of the project. The main key to the solution proved to be communication and adequate governance that involved all stakeholders.

Dike Relocation in the UNESCO-Biosphere Reserve Elbe-Brandenburg River Landscape

The example of the UNESCO-Biosphere Reserve Elbe-Brandenburg River Landscape shows concrete fields of action concerning climate change in protected areas. It names expedient approaches to solve existing conflicts of interest concerning nature conservation, climate change and flood retention. Here a coordinated and balanced management is particularly important. Private landowners, farmers, forestry, municipalities, environmental organisations, flood protection agencies and other (public) authorities form a pool of "experts" for the overriding topic of water retention (flood prevention) measures and for all topics of land use and management. Main challenges in this case are the local socioeconomic conditions and the integration of a variety of stakeholders into a project that is primarily aimed at management objectives represented by the biosphere reserve. The dike relocation was planned as a conservation project and provides, as a "side effect", considerable positive impact on water retention in the biosphere reserve (Gorm et al. 2015).

Brief Description of the Dike Relocation Project

The project area is situated in north-central Germany half-way between Hamburg and Berlin in the biosphere reserve "Flusslandschaft Elbe-Brandenburg", the latter being part of the 400 km biosphere reserve covering five German states along the Elbe River floodplain.

The general idea of the 420-hectare project "dike relocation near Lenzen (Brandenburg)" (Fig. 18.1) was to improve the ecological state of a lowland floodplain, which, over the past centuries, had been turned from a naturally wooded landscape into a mainly agriculturally used landscape dominated by monotonous grasslands. Re-establishing floodplain forests has become an important conservation goal since they are the most species-rich type of forest in central Europe and have become a largely reduced and highly endangered habitat type. Floodplain forests (EU-codes 92E0 und 91F0) are protected by the EU Habitat Directive (European Community 1992). The area is protected by dikes close to the river that largely reduced the floodplain area. In order to reestablish an ecologically functional floodplain, which is primarily an inundatable floodplain, the relocation of the dikes to regain a natural flooding regime was essential. Turning agricultural land back into floodplain forest was a main objective of the project. For this purpose, it was implemented within the federal conservation programme "chance.natur" ('large scale conservation project'), covering 75% of the project expenses. It was furthermore funded by the federal state of Brandenburg—in Germany the states are responsible for conservation issues. The programme requires implementation by a non-governmental organization; accordingly, it was carried out by a local association called "Traegerverbund Burg Lenzen", which is an alliance of the NGO "Friends of the Earth" (BUND) with

Fig. 18.1 Project area in Brandenburg

the local municipality and a number of conservation foundations. This institution already runs a regional environmental education center in the adjacent and historically important castle of Lenzen (Brandenburg)—which also houses a biosphere reserve visitor center.

Availability of suitable land for such large-scale projects is usually a difficult issue. In this case, conflicts over the land use were relatively smoothly solved due to close cooperation of the biosphere reserve administration and the local agricultural holding company, which was the sole tenant of the majority of the agricultural acreage in the project area. Most important, as well as unusual, was the supportive position of the holding company throughout the process. The company's management rated the beneficial effects of the project for the regional development higher than the land loss for the enterprise. Additionally, property issues were solved by a successful 2000-hectare land consolidation scheme that was implemented in order to convert the 420-hectare dike relocation area into public property (Fig. 18.2). A preceding EU-LIFE project targeting the following dike relocation was used to purchase about 550 ha private land (ca. EUR 2 Mio.) spread over a wider region. The land consolidation scheme later swapped this land purchase into the dike relocation area. The tenant was financially compensated for the loss of agricultural area.

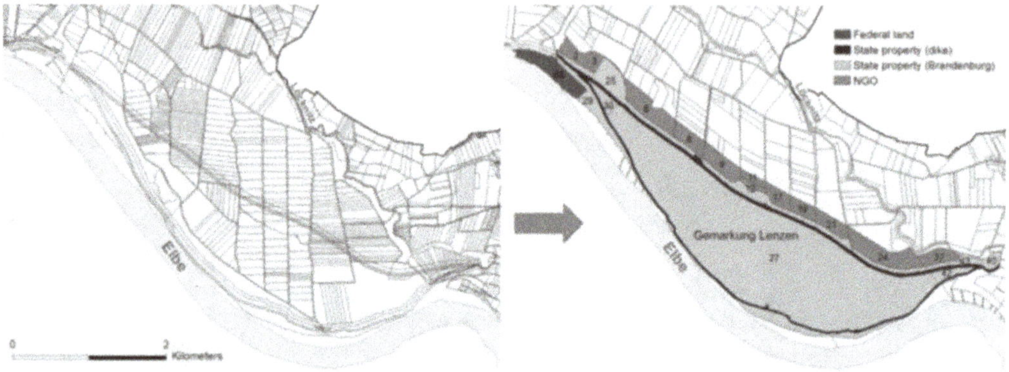

Fig. 18.2 Property situation pre and post project implementation

Stakeholder Process, Project Results and (Public?) Perception

Creating acceptance in the rural environment was of major importance for the project. This was a tedious task concerning the generally difficult economical situation in a remote East German region. A moderation process parallel to the technical planning process was successfully established within the large-scale conservation project. Apart from more than 20 field excursions, nine meetings with representatives of different stakeholder groups and public meetings were held under external moderation. These activities led to an increase of public acceptance and a subsequent planning approval procedure unimpeded by public objections.

Main concerns among local inhabitants were worries about seepage water in housing areas due to a dike line closer to settlements, expected restrictions for hunting and fishing activities and general accessibility of the area. The concerns could mostly be addressed by giving information and a participation process in close connection with the planning process. Scientific evidence was also very helpful in the process: a research program supported by the Federal Ministry for Education and Technology (BMFT) had been carried out before with the intention to assess the options and effects of the dike relocation. The results of this research greatly aided in designing the project and answered many of the questions raised during the moderation phase.

The project attracted national and international attention, as it was the largest dike relocation in Germany at that time. In the beginning, public relations activities focused on a local scale. However, with increasing recognition regionally and beyond, these activities expanded. As a successful pilot project, it proved the beneficial effects of such measures right after its implementation. Main result is the restoration of 420 ha of inundatable floodplain with a mosaic of different floodplain-specific habitat types like wet meadows, reeds, flood channels, softwood and hardwood floodplain forest. Fast successional processes of vegetation and fauna have been observed and are, to some extent, still being monitored. The ecological restoration success was coupled with a considerable effect of flood peak reduction, which was monitored by the Federal Waterways Engineering and Research Institute (Faulhaber 2013) as well as by the Federal Hydrology Institute (Promny et al. 2014). The successful combination of ecological restoration and flood mitigation in particular is widely acknowledged (e.g., European Water Framework Directive 2000/60/EC and Flood Risk Management Directive 2007/60/EC), making the project a blueprint for urgently required water management actions on many other rivers.

An additional result was a positive effect on regional development, for example, in the field of tourism, drawing considerable attention to the region, which is still striving to compensate for the extensive economic losses after the German reunification and the subsequent breakdown of the former socialist economy. The development of its touristic potential, especially eco-tourism, is an explicit objective of the region.

Lessons Learned?

An evaluation of the project was carried out on its technical, conservation-related and social results. As mentioned above, the flood peak reduction was measured during several subsequent flood events as well as a successful reactivation of natural groundwater fluctuations. Considerable increases of populations of birds and fish species have been recorded. On the other hand, reforestation efforts have proven to lag behind expectations, mainly due to the harsh conditions of flooding, drought as well as ice, which plantings have to stand in floodplain situations.

An evaluation of the social environment assessed twice the level of acceptance for the project (Table 18.1). Even though the sample size of the first survey is small, the overall trend of an increase in acceptance becomes obvious.

Although the project's effects toward flood peak reduction reached an extent unprecedented in Germany, addressing climate change as a driver of the project has so far not been an issue. During the planning phase in the early 2000-years, climate change had not been on the agenda in most of Germany, whereas other topics dominated the local and regional discussion of the project. Acceptance among

Table 18.1 Results of a survey on project acceptance within different stakeholder groups: attitude of interviewees towards the conservation project in 2009 compared to the interview in 2004

What is your attitude towards the project (Nov 2009)? (n = 51)	Tenants (n = 4)	Touristic service providers (n = 24)	Other stakeholders (n = 23)	Total	Survey results 2004 (n = 12)
Disapproval	0	1	0	1	4
Mixed feelings	1	5	7	13	5
Indifferent	0	3	0	3	0
Supportive	3	15	16	34	3

(Luley et al. 2010)

the local public was a main issue in the beginning as scepticism mainly toward the dike relocation part of the project ran high. Forest reestablishment and other conservation measures were not criticized as severely. However, a relocated dike, being located more closely to the settlements was largely perceived as an unpredictable threat (Stelzig 2000). Even though scientific data and modelling had demonstrated the measure would result in a significant increase of flood safety by reduced water levels and have only insignificant adverse effects on local hydrological conditions, even local experts remained skeptical for much of the planning phase.

One effect to be considered concerning the background of this hesitant attitude might be seen in the project area's location in eastern Germany, immediately adjacent to the formerly fortified east-western border. The region, of course, has a very special history of limitations and an intense experience of restrictions. A very precautious perception is understandable where large, externally driven projects are seemingly imposed by authorities and not the result of local decisions. One interviewee described this as restrictions formerly imposed by a totalitarian state that will now be imposed by some conservation administration. Widespread prejudices between citizens and actors from East and West Germany, which have been (and to a diminishing extent still are) a side effect of the German unification, also have fostered these conflicts. Given this situation, much of the process concerning the project's contents was not as much a discourse of facts but a projection of societal processes in a region of strong political, economical and societal transitions.

Even though climate change as such had not been addressed specifically, flood protection as a primary reason for the project has been regarded an undisputed asset of the project from the beginning. Increased flood activity has long been known as one of the most easily observed effects of climate change. The considerable effect of this dike relocation on flood peak reduction has been a most convincing argument from the beginning, with particular importance after a catastrophic flood in Eastern Germany in August 2002. Since physical measurements in a number of subsequent flood events furnished this data, the positive effects on flood retention became at least as important for the public perception of the project as the primarily intended ecological improvements. Both issues, flood retention and ecological objectives, have their specific relevance to climate change. In this regard the project's implementation can be interpreted as part of an intended strategy against such developments.

Improving the ecological integrity is considered a conservation strategy in order to increase the resilience of natural systems. The ability of natural systems to withstand disturbance increases with its ecological intactness/state. This applies also to floodplain ecosystems. The well-documented success in species recovery, for example, among wetland bird species already during the implementation of the dike relocation proves a positive effect on species populations. This positive effect is likely to increase their ability to withstand future adverse developments. Thus the implementation of such measures in significant dimensions can be regarded as a strategical mean for an adaptation to climate change on the ecosystem and landscape scale.

Conclusion

Flood prevention in large conservation areas must take into account specific requirements. This applies in particular large-scale structural changes like dike relocations. The described relocation takes place in a region that has to face demographical shrinkage and a lack of economical perspective. Water management as well as natural resource management require an ongoing discussion with private landowners, farmers, public authorities and other stakeholders. As the case study shows, numerous actors with various perspectives have to be involved in decisions, which is an essential element for such complex projects to succeed.

The coexistence of uncertain effects of climate change, various interests of different stakeholders and the requirements of nature protection and sustainable rural development may cause conflicts concerning the need and practical implementation of water retention measures. Within these processes, flood control proves to be a "stronger" aim than nature conservation. The article describes project aims, its implementation in the regional social context and factors considered important for the project success. The case of a dike relocation shows the need for appropriate management to resolve differing demands on land use like agriculture, nature conservation as well as flood protection.

Acknowledgements Open access of this chapter is funded by COST Action No. CA16209 Natural flood retention on private land, LAND4FLOOD (www.land4flood.eu), supported by COST (European Cooperation in Science and Technology).

References

European Community (1992) Council Directive 92/43/EEC of 21 May 1992 on the conservation of natural habitats and of wild fauna and flora. Available via DIALOG. http://eur-lex.europa.eu/legal-content/EN/TXT/?uri=CELEX:31992L0043. Accessed 11 May 2018

Faulhaber P (2013) Zusammenschau und Analyse von Naturmessdaten. In: Die Deichrückverlegung bei Lenzen an der Elbe. BAW-Mitteilungen 97:109–134

Gorm D, Kleeschulte S, Philipsen C, Mysiak J (2015) Exploring nature-based solutions. The role of green infrastructure in mitigating the impacts of weather- and climate change-related natural hazards. EEA Technical report no 12/2015, p 61. ISBN 978-92-9213-693-2. https://doi.org/10.13140/rg.2.2.11273.24169

Hartmann T, Spit T (2016) Legitimizing differentiated flood protection levels—consequences of the European flood risk management plan. Environ Sci Policy 55:361–367. https://doi.org/10.1016/j.envsci.2015.08.013

Yachi S, Loreau M (1999) Biodiversity and ecosystem productivity in a fluctuating environment: the insurance hypothesis. Proc Natl Acad Sci USA 96:1463–1468. https://doi.org/10.1073/pnas.96.4.1463

Kreibich H, Seifert I, Thieken AH, Lindquist E, Wagner K, Merz B (2011) Recent changes in flood preparedness of private households and businesses in Germany. Reg Environ Change 11(1):59–71. https://doi.org/10.1007/s10113-010-0119-3. Available via DIALOG. https://link.springer.com/article/10.1007/s10113-010-0119-3. Accessed 12 Dec 2017

Luley H, Peters J, Christian S, Buss E (2010) Landwirtschaftliche und touristische Nutzungsänderungen im Naturschutzgroßprojekt „Lenzener Elbtalaue" (2005–2009). Sozio-ökonomische Evaluierung (I). Unpublished report, Hochschule Eberswalde

Promny M, Hammer M, Busch N (2014) Untersuchungen zur Wirkung der Deichrückverlegung Lenzen auf das Hochwasser vom Juni 2013 an der unteren Mittelelbe. Korrespondenz Wasserwirtschaft 6(7):344–349

Stelzig I (2000) Akzeptanz von Naturschutzmaßnahmen in Großschutzgebieten—Befragung der Einwohner zweier Dörfer zu Maßnahmen der Auenregeneration. In: Naturbildung und Naturakzeptanz. In: Trommer G, Stelzig I (eds) Frankfurter Beiträge zur biologischen Bildung 2. Shaker Verlag, p 19–46

Warner B, Rannow S (2016) Anpassung an den Klimawandel als Herausforderung für Biosphärenreservate—das Beispiel Flusslandschaft Elbe-Brandenburg. Raumforschung und Raumordnung 74(6):555–567. https://doi.org/10.1007/s13147-016-0425-4. Available via DIALOG. https://link.springer.com/article/10.1007/s13147-016-0425-4. Accessed 12 Dec 2017

Wilke C, Rannow S (2013) Management handbook—a guideline to adapt protected area management to climate change. HABIT-CHANGE report 2013(6). Available via DIALOG. http://www.central2013.eu/fileadmin/user_upload/Downloads/outputlib/HABIT-CHANGE_5_3_2_Management_Handbook.pdf

Dr. Barbara Warner is a social geographer. Since 2014 she has served as head of the academic section Ecology and Landscape of the Academy for Spatial Research and Planning (ARL) in Hanover, Germany. Her thematic work focuses on sustainable land management and sustainable regional development, nature conservation strategies, protection, development of biodiversity and planning in times of climate change and social transformation and transdisciplinary working methods.

Dr. Christian Damm is a floodplain ecologist, since 2009 teaching assistant at Karlsruhe Institute of Technology (KIT), Institute of Geography and Geoecology, Department of Wetland Research, the former WWF-Institute for Floodplain Ecology, Rastatt. Formerly he was project manager of the Large Scale Conservation project "Lenzener Elbtalaue". He is working on river and floodplain restoration, sustainable floodplain management, multifunctional use and ecosystem services of floodplains.

Chapter 19
Commentary: Dike Relocation from an Environmental Policy Perspective

Martijn F. van Staveren

Dike relocation is receiving ample attention in the academic literature, and many detailed case studies have been published on the topic in recent years (Bates and Lund 2013; Eden and Tunstall 2006; Scrase and Sheate 2005; Suddeth 2011; van Staveren et al. 2014; Warner 2008). Germany, the Netherlands and the United Kingdom top the charts of dike relocation projects in Europe, while also many regional programmes, such as in the vast Danube delta area, provide additional examples. Outside Europe, notably in the United States, large stretches of levees (synonym to dikes or embankments) in the Sacramento-San Joaquin delta in California, as well as along the Mississippi and its tributaries, have been relocated to increase rivers' discharge capacities, to facilitate flooding on previously enclosed land, and to restore overall natural dynamics in the floodplain.

The dike relocation case in the River Landscape Elbe-Brandenburg biosphere reserve presents interesting and useful insights into the emergence and apparent popularity of dike relocation initiatives; a shared objective with the studies referred to above. A key driver in this particular case was the governmental effort to turn agricultural land into floodplain forest, with the objective to improve the ecological state of the floodplain. For a deeper understanding of this driver, we can learn from other studies: calls for environmental protection and environmental movements since the 1970s strongly influenced water policy, where eco-inspired minds advocated for new ways of integrated environmental management, including water and ecology (Disco 2002; Saeijs 2008). This was the conceptual foundation of various approaches to flood management, from Building with Nature (Waterman 2008) to, most recently, Nature-based solutions and Ecosystem-based Disaster Risk Reduction (Renaud et al. 2013).

Similar to other dike relocation projects, the Elbe-Brandenburg project clearly identifies the governmental authorities' argument that rivers need more space, in

M. F. van Staveren (✉)
Environmental Policy Group, Department of Social Sciences, Wageningen University, Wageningen, Netherlands
e-mail: martijn.vanstaveren@wur.nl

T. Hartmann et al. (eds.), *Nature-Based Flood Risk Management on Private Land*,
https://doi.org/10.1007/978-3-030-23842-1_19

order to facilitate natural dynamics (flooding, nature restoration) to freely take their course. But after detailed investigation, such projects often display tight forms of flood control. Cases in the Netherlands show that it is meticulously predicted and controlled when, where and how much flooding is allowed on reconnected floodplains (van Staveren et al. 2014). Vegetation growth in widened floodplains has to be kept within certain physical boundaries in order to avoid obstructing overland flood conveyance. Some authors have therefore stated that room for the river is a synonym for "room for the engineer" (Van Hemert 1999). A question that also applies in the case of the River Landscape Elbe-Brandenburg biosphere reserve is to which extent dike relocations can be genuinely labelled as NBS, and to which extent they are "camouflaged" varieties of hydrocratic flood control.

Furthermore, it is useful to make a distinction between dike relocation projects along rivers on one hand and in coastal zones on the other. The geographical setting, providing terms of reference for environmental dynamics, determines the spatial possibilities for dike relocations. Besides the obvious variations in water quality (fresh water in the riverine area, and brackish or saline water in the coastal zone) influencing what kind of nature could be restored, also water-related dynamics strongly differ. Water dynamics in riverine areas are strongly season-based, which usually means that winter seasons come with higher and peak discharges compared to drier summers. In the coastal zone, water dynamics are influenced by river flows but also by twice-daily tides, pushing saline water into the estuary. Bodies of literature on both types of dike relocation have emerged (Borsje et al. 2011; French 2006; Warner et al. 2013; Waterman 2008). Each present a different terminology and concepts, such as "managed coastal realignment" (French 2006) and "Space for the River" (Warner et al. 2013).

In scientific disciplinary terms, the Elbe-Brandenburg case touches on policy studies, governance issues and stakeholder participation. Stakeholder participation in water management is a well-studied topic (Warner et al. 2013), and a shift from government to governance is indeed a key driver for more inclusive stakeholder participation projects in water management. A key challenge for policy makers here is to find a right balance between the extent of stakeholder involvement, and making decisions that might negatively impact some of these actors. In the dike relocation case in the River Landscape Elbe-Brandenburg biosphere reserve, stakeholders were kept onboard by means of structural involvement, not only with paper plans but also via a series of funded projects.

From the perspective of policy research, many dike relocation studies, including the piece under scrutiny, share the observation that dike relocations no longer concern a single policy domain. Nature restoration and flood prevention initiatives are often encountered simultaneously, leading some authors to speak of "green-blue alliances" when it comes to pursuing policy implementation in river management. In the Netherlands in particular, dike modifications date back to calls for environmental protection and restoration in the 1970s; the time of growing environmental concerns and environmental movements. This went hand in hand with critique on governments pursuing large-scale infrastructure with potentially negative environmental impacts. A third policy domain, that of climate change adaptation, became intertwined with nature restoration and water management after extensive (near) flooding in the mid-1990s in Germany and the Netherlands presented an impetus to rethink river management policies over longer timescales (Van Heezik 2008). It is relevant to highlight, as the dike relocation case in the River Landscape Elbe-Brandenburg biosphere reserve also brings forward, that flood control is a stronger aim than nature restoration. The fourth strongly related policy domain, agriculture, often gets the worst of it in the shape of loss of (privately owned) land.

It is important to be careful to interpret the popularity of dike relocations being representational for a policy transition, which usually implies a shift from one style of river management to a different one. This would be tempting, as dike relocations and river widening ("opening up") fully contrasts with the conventional approach ("closing off") (Van Staveren et al. 2017). Transitions are often presented as a shift from approach A to B, where B supersedes the former approach A. However, comparing the length or surface area of dike relocations and river widening, it often turns out that they are very limited compared to modifications done to the remaining dike system. Even more, in the Netherlands, a vast amount of 943 km of dikes will be strengthened predominantly by means of "traditional" techniques to increase height and install reinforcements by 2050 in order to meet more strict flood safety norms (Ministry of Infrastructure and Water Management 2019). As mentioned earlier, dike relocation projects carry with them strong notions of flood control. River restoration in the United States receives ample attention and substantial financial investment these days (The Nature Conservancy 2018). Although it is obvious that FRM policies have changed over the last decades, dike relocation initiatives can best be regarded as taking place on the fringes of dominant full flood prevention approaches in FRM. In policy studies vocabulary, they might be classified as niche developments (Geels and Schot 2007).

The key finding of the dike relocation case in the River Landscape Elbe-Brandenburg biosphere reserve is that numerous actors with various perspectives have to be involved in decision making (Warner and Damm, this volume). This is certainly a challenge for policy implementers, who are often confronted with a large (local) stakeholder group, different opinions and various policy domains that have to be incorporated in final project designs. In that sense, dike relocations in areas with a high population density and intensive forms of land use have additional challenges. The implementation of the dike relocation case in the River Landscape Elbe-Brandenburg biosphere reserve benefited from that fact that it is a "remote region" with "the lowest population density in all of Germany" (Ministerium für Umwelt Gesundheit und Verbraucherschutz 2013) and a "difficult economic situation" against the background of the East-West divide and unification while substantial budgets were available from an external funder (Warner and Damm, this volume). These are important contextual factors in support of project acceptance and implementation. It is interesting to note that the authors argue that the earlier experience with a totalitarian state in the region would foster resistance to restoration efforts, rather than easy acceptance of state-driven projects (see on the topic also Scott 2008).

Finally, I relate their findings to the general objectives of the book on NBS. The authors underscore both challenges and stimulating factors related to dike relocation projects as examples of NBS. Integrated thinking about river management can be seen as a no regret approach; whether projects are initiated from nature restoration or from a flood management perspective, integrated approaches stimulate the weaving together of policy domains in search of multifunctional interventions and win-win situations. On the other hand, it is also worthwhile to be realistic about the potential of dike relocations as examples of NBS. They can be suitable for some geographical settings, but will be more difficult to implement in areas with high population densities and intensive forms of land use.

Acknowledgements Open access of this chapter is funded by COST Action No. CA16209 Natural flood retention on private land, LAND4FLOOD (www.land4flood.eu), supported by COST (European Cooperation in Science and Technology).

References

Bates ME, Lund JR (2013) Delta subsidence reversal, levee failure, and aquatic habitat—a cautionary tale. San Francisco Estuary and Watershed Science. DIALOG. https://escholarship.org/uc/item/9pp3n639

Borsje BW, van Wesenbeeck BK, Dekker F, Paalvast P, Bouma TJ, van Katwijk MM, de Vries MB (2011) How ecological engineering can serve in coastal protection. Ecol Eng 37(1):113–122. https://doi.org/10.1016/j.ecoleng.2010.11.027

Disco C (2002) Remaking "nature": the ecological turn in Dutch water management. Sci Technol Hum Values 27(2):206–235. https://doi.org/10.1177/016224390202700202

Eden SE, Tunstall S (2006) Ecological versus social restoration? how urban river restoration challenges but also fails to challenge the science-policy nexus in the United Kingdom. Environ Plann C Gov Policy 24(5):661–680. https://doi.org/10.1068/c0608j

French PW (2006) Managed realignment—the developing story of a comparatively new approach to soft engineering. Coast Shelf Sci Estuar 67(3):409–423. https://doi.org/10.1016/j.ecss.2005.11.035

Geels FW, Schot JW (2007) Typology of sociotechnical transition pathways. Res Policy 36(3):399–417. https://doi.org/10.1016/j.respol.2007.01.003

Ministerium für Umwelt Gesundheit und Verbraucherschutz (2013) Elbe-Brandenburg river landscape biosphere reserve. https://www.elbe-brandenburg-biosphaerenreservat.de/fileadmin/user_upload/PDF/LfU/Gebietsfaltblaetter_englisch/br_elbe_eng.pdf

Ministry of Infrastructure and Water Management (2019) Delta programme 2019. https://english.deltacommissaris.nl/delta-programme/documents/publications/2018/09/18/dp2019-en-printversie

Renaud FG, Sudmeier-Rieux K, Estrella M (eds) (2013) The role of ecosystems in disaster risk reduction. United Nations University Press, Bonn

Saeijs HLF (2008) Turning the tide. Delft Academic Press, Delft, Essays on Dutch ways with water

Scott JD (2008) Seeing like a state. Yale University Press, Yale

Scrase JI, Sheate WR (2005) Re-framing flood control in England and Wales. Environ Values 14(1)113–137. http://sro.sussex.ac.uk/26165/

Suddeth R (2011) Policy implications of permanently flooded Islands in the Sacramento–San Joaquin Delta. San Francisco Estuary Watershed Sci 9(2):1–18. http://doi.org/10.15447/sfews.2011v9iss2art5

The Nature Conservancy (2018) New river plan will guide next decade of restoration along the Hudson. https://www.nature.org/en-us/explore/newsroom/new-river-plan-will-guide-next-decade-of-restoration-along-the-h/. Accessed 25 Nov 2018

Van Heezik A (2008) Battle over the rivers. Two hundred years of river policy in the Netherlands, VHB and Rijkswaterstaat, Haarlem

Van Hemert M (1999) Ruimte voor de ingenieur. Rivierbeheer in Nederland eind jaren negentig. K&M, Tijdschrift voor Empirische Filosofie 361–387

Van Staveren MF, Warner JF, Khan MSA (2017) Bringing in the tides. From closing down to opening up delta polders via Tidal River Management in the southwest delta of Bangladesh. Water Policy 19(1). http://doi.org/10.2166/wp.2016.029

van Staveren MF, Warner JF, van Tatenhove JPM, Wester P (2014) Let's bring in the floods: de-poldering in the Netherlands as a strategy for long-term delta survival? Water Int 39(5). http://doi.org/10.1080/02508060.2014.957510

Warner JF (2008) Emergency river storage in the Ooij polder—a bridge too far? Forms of participation in flood preparedness policy. Int J Water Resour Dev 24(4):567–582. https://doi.org/10.1080/07900620801923153

Warner JF, Van Buuren A, Edelenbos J (2013) Making space for the river. Governance experiences with multifunctional river flood management in the US and Europe. IWA Publishing, London

Waterman RE (2008) Integrated coastal policy via building with nature. Dissertation, Delft University of Technology, Delft

Martijn van Staveren works at the Environmental Policy Group at Wageningen University. He completed his Ph.D., a comparative study on flood management policy in three different countries, in 2017 and continued as a postdoctoral researcher in a new research programme concerning the implementation of new flood safety norms in the Netherlands. His thematic focus is on the socio-political dimensions and knowledge co-development of flood risk management, with an interest in achieving long-term resilience to flooding and exploring the opportunities of ecosystem-based flood risk management.

Chapter 20
Commentary: Reflection on Governance Challenges in Large-Scale River Restoration Actions

Kristina Veidemane

Strong stakeholder and actor engagement in the development of diverse environment and nature-related policies are an important component to ensure successful design and implementation of measures for achieving the set policy objectives. Many social methods are tested and applied in policy planning; however, the key issue is still to explore the most adequate and successful approaches and methods for those policies that require multi-scalar and cross-sectoral approaches for the stakeholder involvement processes, like water or FRM policies.

River ecosystems provide multiple benefits for society, defined also as ecosystem services. The European Union Biodiversity Strategy adopted in 2010 has determined Task 2 to be to maintain and restore ecosystems and their services by including green infrastructure in spatial planning and restoring at least 15% of degraded ecosystems by 2020. The objective of the EU Floods Directive adopted in 2007 is the establishment of a framework for measures to reduce risks of flood damage, including flood prevention, protection and preparedness. The Floods Directive mentions that a flood management plan shall consider where possible the maintenance and/or restoration of floodplains, as well as measures to prevent and reduce damage to human health, the environment, cultural heritage and economic activity. Although, there is an interlink between both policy objectives, a key challenge is how EU policies are implemented via planning and management schemes at national, regional and local levels, respecting ecological, social and economic conditions.

Nowadays spatial, land use or development planning can no longer be performed without the involvement of society and different stakeholder groups. This is ensured by the Aarhus Convention (UNECE 1998), which established a right of the public to participate in environmental decision-making. People have a right to express their opinion or comment; consequently; they have an opportunity to influence decision-making that might be driven by economic profits and unfavourable for nature conservation or quality of life. A number of cases have been collected and presented

K. Veidemane (✉)
Baltic Environmental Forum (BEF), Riga, Latvia
e-mail: Kristina.Veidemane@bef.lv

T. Hartmann et al. (eds.), *Nature-Based Flood Risk Management on Private Land*,
https://doi.org/10.1007/978-3-030-23842-1_20

by the UNECE Aarhus CLEARINGHOUSE for Environmental Democracy (Justice and Environment—European Network of Environment Law Organizations 2013). In order to achieve optimal planning solutions, various biophysical, social and economic methods shall be applied depending on the scale and context of the project described by Warner and Damm (this volume). A scientific approach to the problem solving, evidence and knowledge-based planning process are acknowledged and utilised more frequently by practitioners.

The relocation of the dike on the river Elbe (near Lenzen, Brandenburg) presents the project that was launched in the 2000s before the adoption of the EU Biodiversity Strategy 2020 targets; the Floods Directive was in the drafting stage then. The project was definitely innovative in those days and has provided an input for other locations, for example, in Germany and Poland. Therefore, the experiences and knowledge gained during such a project as the dike relocation at the Elbe is rather unique and valuable for all stakeholders. Ex-post monitoring of nature and water status provides evidence for the effectiveness of such measures in the short and long term respectively and thus potential for its replication.

Driven by nature conservation policy, in particularly, restoration of floodplains, relocation of a dike causes significant impact on water resources, which demands a complex and integrated planning approach. Floodplain restoration improves river hydro-morphology, which is one of the key components to assess the water status according to the EU Water Framework Directive. Recognising synergies and multiple benefits between nature conservation policies, flood protection policies and socio-economic benefits is a very important success factor in achieving desired goals. Several studies have already addressed the synergies between the nature, water and flood risk policies of the EU to achieve the best for rivers and people (EEA 2015; Evers 2016; Ignar and Grygoruk 2015; Schindler et al. 2016). However, less attention is paid to whether and how the EU and national policy objectives are shared at the regional and local level. Are there synergies or conflicts between the governmental levels and how research can support to achieve policy coherence?

Multifunctional landscapes represent also a higher degree of diversity among stakeholders to be involved in policy planning and implementation process. The case area represents rather monofunctional agricultural landscape dominated by grasslands. The key stakeholder was one local agricultural holding company managing almost the entire case study area and demonstrated a positive attitude towards the initiative. Moreover, its loss was compensated financially. In the majority of planning cases, fragmented patterns of land ownership present notable challenges for policy makers and constraints for enhancing development towards NBS at a larger scale (Kabisch et al. 2016; Wamsler et al. 2017). How much of the success of a proposed green infrastructure project depends on the method selected to run the stakeholder process? A variety of tools and techniques are available, but which are most effective and flexible? Can communication and awareness raising activities deliver landowner acceptance? The project was implemented in the protected area—a biosphere reserve where people are used to certain demands imposed by nature conservation. Is this one of the success factors for selecting a site for innovative measures?

The case highlights the importance of the overall acceptance of a large-scale infrastructure project, as it encompasses not only farmland but creates change in the rural landscape. Therefore, stakeholder involvement and communication were organised beyond the group of directly impacted stakeholders, such as landowners whose estates are in the main focus. A survey carried out to assess stakeholders' attitude towards the project illustrates an increasing acceptance of the project. Amongst others, the tourism sector has been identified as one of the stakeholder groups in the project area assessing the project positively.

Data and hydrological models and involvement of research and monitoring institutes to support the planning process with evidences about the impact on water levels in the local area was another core of the project. Monitoring effects after the reallocation of the dike proved that the peak flows are reduced and water is retained in the space allocated for the river. Nowadays, effectiveness of the flood retention or river restoration measures is monitored and assessed as the restoration projects are mainly financed by public funding. For example, the project co-financed by the EU LIFE Programme 2014-2020 shall implement monitoring activities to assess the impact of implemented actions (Regulation [EU] No. 1293/2013).

Water retention measures are gradually implemented at various scales and deliver numerous social, economic and environmental benefits. Assessing and valuing multiple benefits requires diverse and flexible methodologies from different disciplines; thus research teams shall be built to cover the competencies and skills of a interdisciplinary nature surrounding the topic. Moreover, the assessment results concerning the benefits shall be presented in targeted and tailored ways as different stakeholder groups have their own value systems. Sharing policy goals between various government layers and stakeholder groups is not self-evident; therefore it is important to collaborate across levels and interest groups.

Acknowledgements Open access of this chapter is funded by COST Action No. CA16209 Natural flood retention on private land, LAND4FLOOD (www.land4flood.eu), supported by COST (European Cooperation in Science and Technology).

References

European Environment Agency (EEA) (2015) Flood risks and environmental vulnerability Exploring the synergies between floodplain restoration, water policies and thematic policies, p 78

Evers M (2016) Integrative river basin management: challenges and methodologies within the German planning system. Environ Earth Sci 75:1085. https://doi.org/10.1007/s12665-016-5871-3

Ignar S, Grygoruk M (eds) (2015) Wetlands and water framework directive protection, management and climate change. https://doi.org/10.1007/978-3-319-13764-3

Justice and Environment—European Network of Environment Law Organizations (2013) Public participation in spatial planning procedures: comparative study of six EU member states. https://aarhusclearinghouse.unece.org/resources. Accessed 28 Dec 2018

Kabisch N, Frantzeskaki N, Pauleit S, Naumann S, Davis M, Artmann M, Zaunberger K (2016) Nature-based solutions to climate change mitigation and adaptation in urban areas: perspectives on indicators, knowledge gaps, barriers, and opportunities for action. Ecol Soc 21(2):39

Regulation (EU) No 1293 (2013) of the European Parliament and of the Council of 11 December 2013 on the establishment of a Programme for the Environment and Climate Action (LIFE) and repealing Regulation (EC) No 614/2007 Text with EEA relevance. OJ of the EU. L 347:185–208

Schindler S, O'Neill FH, Biró M et al (2016) Multifunctional floodplain management and biodiversity effects: a knowledge synthesis for six European countries. Biodivers Conserv 25(7):1349–1382. https://doi.org/10.1007/s10531-016-1129-3

United Nations Economic Commission for Europe (1998) Convention on access to information, public participation in decision-making and access to justice in environmental matters (Aarhus convention). Available via DIALOG. http://www.unece.org/env/pp/treatytext.html

Wamsler C, Pauleit S, Zölch T, Schetke S, Mascarenhas A (2017) Mainstreaming nature-based solutions for climate change adaptation in urban governance and planning. In: Nature-based solutions to climate change adaptation in urban areas. Springer, Cham, pp 257–273

Dr. Kristina Veidemane has a background in geography and environmental management. She has more than 20 years of working experience in the field. She is leading water management and stakeholder involvement expert in the Baltic Environmental Forum. Since the beginning of the 2000s, Dr. Veidemane has been active in promoting cross-border cooperation and implementing measures aiming at improving water management of transboundary rivers in the Baltic States.

Part VIII
Multi-Use Climate Adaptation in the Netherlands

Chapter 21
Dilemmas of an Integrated Multi-use Climate Adaptation Project in the Netherlands: The Oekense Beek

Maria Kaufmann and Mark Wiering

This case study presents an integrated, collaborative and multi-use climate change adaptation project in the Netherlands: the "*Oekense Beek*". The aim of this chapter is to illustrate the challenges of such a project in a context where the cooperation with private landowners and users is mainly based on a voluntary approach. Governmental authorities have several policy instruments at their disposal that may help them find "agreements" with private landowners. However, most agreements need to be found on a voluntary base with financial incentives or communicative approaches. Consequently, governmental authorities are confronted with several challenges, such as the lack of urgency and awareness for climate change among stakeholders, the long duration and iterative adjustment of project plans due to an increasing number of involved stakeholders, coordinating multiple land uses, developing innovative synergies and ensuring an equal treatment of land users. These aspects make it difficult to set up and implement an integrated climate change adaptation project.

Introducing the Climate Change Adaptation Project *Oekense Beek*—An Integrated Approach

Climate change is often described as a wicked problem (Termeer et al. 2013), which asks for new forms of governance based on the coordinated efforts of various governmental and non-governmental stakeholders from different policy sectors (Olsson et al. 2006; Ansell and Gash 2007; Collins and Ison 2009; Baird et al. 2016). However, the complexities of such integrated and non-hierarchical decision-making processes tend to be underestimated (Wise et al. 2014; Biesbroek et al. 2015; Duit 2015; Sjöst-

M. Kaufmann (✉) · M. Wiering
Institute for Management Research, Radboud University Nijmegen, Nijmegen, Netherlands
e-mail: m.kaufmann@fm.ru.nl

M. Wiering
e-mail: m.wiering@fm.ru.nl

T. Hartmann et al. (eds.), *Nature-Based Flood Risk Management on Private Land*,
https://doi.org/10.1007/978-3-030-23842-1_21

edt 2015), particularly in a context where private landowners and users are involved. This case study presents an integrated, collaborative and multi-use climate change adaptation project in the Netherlands. The aim of this chapter is to illustrate the challenges of such a project in a context where the cooperation of private landowners and users is given on a voluntary basis.

The climate change adaptation project "*Oekense Beek*" is an interesting case as it actively aims for an integrated approach. The project aims to go beyond the traditional distinctions made in water quantity management. It aims to manage floods and droughts in an integrated manner, acknowledging the disparate consequences of climate change, which may increase the intensity of both flood events and dry periods. In this project, the solutions to these issues are found in nature management, for example, recovery of natural hydrological dynamics or improving the quality of the soil to increase the water infiltration capacity. A representative of the province described the vision of the project to connect sectorial approaches. The representatives hope that this project may become a pilot for the East of the Netherlands, where several areas may face similar climate change consequences.

Apart from involving various policy sectors, cooperation with private landowners and users is an essential element for realizing the project as measures have to be implemented on private land. Multiple forms of land-use need to be coordinated. The main landowners in the project area are a nature organisation—*Natuurmonumenten*—and farmers with crop fields and livestock. The chapter addresses following questions: How do governmental authorities approach private landowners and users? What kinds of instruments have governmental authorities at their disposal to find agreements with private landowners in the context of climate change adaptation? What are the challenges and dilemmas of such an integrated and collaborative climate change adaptation project? At the moment, the project is delayed and remains in the development phase. Due to the difficulties in finding broad support for the project among the local stakeholders, the next steps will be to carefully map the current status and identify what is actually possible to increase climate resilience. Regional authorities think that this will facilitate communication and collaboration among local stakeholders.

Data Collection and Analysis

Inspired by the literature on governance and policy instruments (Arts et al. 2006; Bemelmans-Videc et al. 2010; Driessen et al. 2012), this chapter focuses on the involved actors, their interests and how they communicate and cooperate within the project. The chapter explores the procedural governance approach and the policy instruments governmental officials have at their disposal to find agreements with private land users or owners: coercive instruments ("sticks"), financial incentives ("carrots") and communicative instrument ("sermons") (see Bemelmans-Videc et al. 2010). The chapter is based on interviews conducted in autumn 2017 with governmental representatives: the current project manager from the regional water authority

Vallei en Veluwe, the initiator of the project from the province of Gelderland and a telephone interview with a policy maker of the municipality Brummen. Additionally, policy documents and research reports have been analysed. The documents and the transcribed interviews were analysed deductively according to the following criteria: stakeholders and tasks, policy instruments and barriers.

Setting the Scene: Locating the *Oekense Beek*

The project is located in the eastern part of the Netherlands in the province of Gelderland. The next small-sized city is Zutphen (47,000 inhabitants in 2017). The project is situated in the transition zone between the *Veluwe*, a hilly area with forests and sand soil, and the Ijssel, a river. Both the *Veluwe* and the Ijsseldelta are Natura 2000 areas and part of a larger ecological corridor. The project covers the complete stream *Oekense Beek*, which rises close to the *Veluwe* and discharges into the Ijssel, just like the streams: *Rhienderense Beek* and the *Voorstondese Beek*. The landscape is very diverse, a result of the country estates of the 14th and 15th century. It is characterised by a mix of small-scale forests, streams, fields and moor (Interviews 2017; Provincie Gelderland 2016).

The *Oekense Beek* is a straightened and deep stream. The area is strongly drained. Consequently, the groundwater levels are rather low, which lowers the water infiltration capacity of the area and diminishes its buffer function. Consequently, the stream responds very quickly to precipitation events, or the lack thereof. A representative of the regional water authority describes the stream as follows: "When I visited the stream in early spring this year, when it was still raining, the stream was completely filled with water. When I came along in May, the stream lay dry, the same in August. In September, it started to carry a little bit water." With regards to climate change, this may lead to increasing floods in the future as the intensity and frequency of precipitation events is projected to increase. But also droughts are projected to increase in the future (Interview 2017).

To combat these projections, the aim of the project is to increase the buffer capacity of the area so that the increase in precipitation can be compensated. In other words, the sponge function of the stream shall be increased so that an excess of water can be retained during times of intense rainfall discharging it slower and storing it for times of drought. This includes a number of measures that will promote a better infiltration of water in the area and a slower discharge of water so that the water can also be stored for periods of droughts. Measures include for example, the re-naturalisation of the stream by recovering its old meanders with broad shores and pools that can be inundated. These meanders are still visible in the landscape and can be reconnected. Through these measures the groundwater level will be raised. This is also needed to achieve the Natura 2000 objectives in the area, which demand, for example, the development of floodplain vegetation (*natte natuur*) and alluvial forest. Other climate change adaptation measures include improving the quality of the soil. For example, a higher concentration of organic substances in the soil increases the soil's water infiltration capacity (Provincie Gelderland 2012; Stroming 2013; Interviews 2017).

Who Is Who? Mapping the Actor Landscape

In the Netherlands, the distribution of responsibilities and competencies is quite clearly structured. In the climate adaptation project *Oekense Beek*, the main actors involved are the province of Gelderland, the regional water authority *Vallei en Veluwe*, the municipality Brummen, *natuurmonumenten* and farmers.

The Province of Gelderland: An Integral Organisation

The province of Gelderland is one of 12 provinces in the Netherlands. Dutch provinces form a regional generic administrative layer between the national government and municipalities. They are responsible for spatial planning. In so-called "structural visions", they designate areas for residential housing, economic purposes, nature development, agricultural activities, recreation, traffic, etc. In other words, they take on an integrated perspective considering various interests. Provinces often take on the role of an area manager cooperating closely with different governmental and non-governmental actors (Kaufmann et al. 2016).

In this project, the province initiated the plan but delegated the preparation, planning and implementation to the regional water authority who took over the role of area manager. This decision was based on two main reasons; firstly, water plays a fundamental role in this project: achieving the objectives asks for fundamental changes in the water system. Secondly, the project asks a lot of contact with local users: *natuurmonumenten* and farmers. As the regional water authorities traditionally have more contact with these actors, they have more experience with establishing contact with them. Delegating the responsibility for projects is a common practice in the province (Interviews 2017).

The provincial responsible was very active in linking this project up to various funds to make subsidies available for the different measures (see below). As he put it: "I look around in that organization [province] and see where money is available. You tell a nice story so that the different funds could all invest into your project. You can make a nice start with that collected money" (Interview 2017). This indicates that the pro-activeness and networking capabilities of the involved actors are important to develop and finance such an integrated project.

The Regional Water Authority "Vallei en Veluwe"

In the Netherlands, regional water authorities (a sector-based layer of regional government, with independent tasks and tasks delegated from the provinces) are democratically elected organisations responsible, among others, for water level management (e.g., operating pumps, or maintaining primary and regional flood defences) (Kaufmann et al. 2016). In this project, *Vallei en Veluwe* took on, as explained above, the role of the area manager. It is responsible for preparing and implementing the climate change adaptation measures. A representative of the water authority summarized the main interest of the authority as follows: "If we can't guarantee dry feet for the people living in the area, then our administrative board will not cooperate." The regional water authority is the owner and manager of the water course. *Vallei en Veluwe* divides its regional water courses into different categories: A, B and C. The *Oekense Beek* is an A-water course, which means the regional water authority is responsible for its management. In contrast, B- and C-water courses are the responsibility of the owner. B-water courses are reviewed twice per year by the regional water authority (Interviews 2017).

In the Netherlands, the regional water authorities used to be very autonomous actors focused on their hydro-engineering expertise to take decisions based on quantitative calculations and models. In the last decades, the organisations have become increasingly integrated, taking into account other values and interests (Kaufmann et al. 2016). How much progress has been made in terms of this developments differs among the 22 regional water authorities. Even though '*Vallei en Veluwe*' is described as a relatively integrated organisation, representatives of other governmental organisations also state that there is still room for improvement in terms of increasing the "integral thinking" (Interviews 2017).

As an area manager, the authority is also responsible for approaching the different actors, including private landowners, in this area. When approaching private actors, the representative of the regional water authority takes on an integrated position representing all involved governmental and non-governmental actors, namely the province, municipality Brummen, but also the *natuurmonumenten*. The representative of the regional water authority is hired by, for example, the municipality to represent their interests. This has the advantage of giving the citizens a clear contact person (Interviews 2017).

The Municipality of Brummen

In the Netherlands, municipalities are responsible for developing a local land use plan, where they designate particular land-uses (Kaufmann et al. 2016). The designation of plans cannot so easily be changed from, for example, farmland to nature. If the designation is changed, it needs to be compensated. The project takes place in the municipality Brummen. The municipality appears to be enthusiastic and interested in the project. It does not take a leading role but is interested in creating a liveable area for its citizens, particularly with regards to recreation. It provides financial means to develop recreational opportunities in the project area (Interviews 2017).

Natuurmonumenten

Natuurmonumenten is a non-governmental nature conservation organisation, which is above all interested in nature development: natural processes and connectivity between nature areas. Its inclination toward the project is positive and supportive. The NGO owns land in the area, for example, the old country estates *Leusveld* and *Voorstonden*. The designation of this land is nature function and partly Natura 2000 areas. *Natuurmonumenten* started already with implementing measures on their land. They started logging trees to revive the meander (Interviews 2017). *Natuurmonumenten* leases part of their land to farmers. These are mainly long-term leasehold contracts (*erfpachtcontracten*) (Provincie Gelderland 2016).

Farmers

The project illustrates that farmers are by no means a homogeneous group but have heterogeneous interests and approaches. The activities carried out in the project, for example, raising the water level, are, of course, not necessarily beneficial for farming activities. The periods when the soil is wet may increase from two to four months. According to the interviewees, many farmers argue that they need to prioritise running a business and making money, a concern that appears to be taken seriously by the provincial, municipal and water authority representatives. One organic farmer leases land from *natuurmonumenten*. Even he is not interested in measures at the moment as he wants to keep his livestock outside on the fields, which is not possible when they are too wet. On the other hand, one farmer fundamentally changed course: he sold his livestock and adopted a more sustainable farming practice. He is now nature conservation manager for *natuurmonumenten* and plants particular types of vegetation. He receives financial payment for delivering these green services. The governmental representatives would like to use his story as an exemplary illustration for other farmers and encourage him to become an ambassador for the area (Interviews 2017).

Developing an Integrated Cooperation with Governmental and Non-governmental Actors

The project was initiated by the province of Gelderland. The initial aim was to address the nature development objectives in the context of Natura2000. These objectives included connecting various ecosystems and developing alluvial forest, among others. As the province was also actively involved in the discussions surrounding the Western European Climate Corridor, the project was combined with climate change adaptation initiatives (Interviews 2017).

The province developed a broad "umbrella vision" for the area. Subsequently, it analysed which actors could play a role in this area and discussed internally what could be their interests in the project. Afterwards the province presented their umbrella vision to the other actors to discuss it collectively. In this case, these actors included the regional water authority, the municipality Brummen, *natuurmonumenten*, and LTO. In these discussions, the province tried not to take a dominant or authoritarian position but aimed to have an open deliberation. Several interviewees pointed out that emphasizing the interests of different parties is very important in such a project. As one provincial respondent described his approach, "You get together and say: We have this broad idea. What do you think about it? What are your interests? Where do you see problem? How could we address your problems so that we can cooperate in this project?" The main actors—meaning the province, regional water authority, municipality Brummen, *natuurmonumenten*—agreed on the aim of the project in terms of climate change adaptation and nature development. Together, they deliberated about potential measures and strategies (Interviews 2017).

As these objectives asked for fundamental changes in the water system, several hydrological studies were conducted to model the surface water and the groundwater of the area to get a good picture of the water system. These models considered also various climate change scenarios. Additionally, geological and soil characteristics were analysed. Scientific experts and analysts as well as the governmental authorities come together in expert meetings and combine the different maps to explore which measures are actually feasible. This process is still in progress. A representative of the regional water authority mentioned that it is important to know the range of possible measures and actions. Otherwise, it is difficult to go into negotiations with private stakeholders because the governmental representatives do not know whether they can accommodate the wishes of stakeholders. Based on these models, it will be decided which plots of land are really needed for the climate change adaptation measures. Then the project manager can reach out to the landowner and see whether the land may be bought or other forms of agreements may be found (Interviews 2017).

Finding Agreements with Private Landowners or Users: Available Instruments

Governmental authorities have a number of instruments at their disposal to find agreements with private landowners or users, either in the form of acquiring privately owned land or by convincing private users or landowners to cooperate.

Communicative Instruments—Starting the Conversation with Private Landowners and Users

Getting into contact with private landowners and users and talking about the ideas was the basis of this project. It was necessary to make contact with private landowners or landusers early on as the gauging network needed to be extended to collect the necessary scientific data. These gauging units are placed on private land. Obviously, the owners wanted to know why gauging units were placed in their gardens; therefore they were informed about the preliminary plans. Already in 2013, a number of so-called explorative kitchen table talks had been conducted with about 85 private stakeholders from the area. Stakeholders appeared to experience these talks rather positively. The talks generated trust, and people felt taken seriously as they were ensured that their knowledge and input would be considered. The stakeholders thought that this approach would facilitate collaborative thinking, where a lot of different perspectives would be considered and old paradigms could be overcome (Landschapsnetwerk Brummen 2013). According to an internal background report of the representative of the former DLG (*Dienst Landelijk Gebied*), a majority of the local people were enthusiastic about the project and the idea (Interviews 2017).

Back then the talks were conducted by the former DLG (*Dienst Landelijk Gebied*), which was part of the province. This governmental authority supported the development of rural areas. The main tasks included acquiring lands, reorganizing them, advising on their management and transferring them to area managers or individual farmers. In 2015, the DLG was closed, which was experienced as a great loss. Other individuals from other governmental authorities had to take over the conversations with the stakeholders, which caused delay and made the negotiations with stakeholders more difficult (Interviews 2017).

Nowadays, a presentative of the regional water authority has taken up the talks again. He has talked with 40–50 people during the last two years (2016–2017). The governmental representatives stress that they need and want the input of the local stakeholders. They want to understand citizens' problems, fears and wishes. By posing the right questions, governmental authorities can understand why citizens are doubtful, where particular knowledge or insights are missing. That enables them to develop solutions that also consider the citizens' position. It is possible to stimulate citizens and develop links to their own interests. Additionally, they set up a website to inform about the progress of the project. The website is regularly updated so that citizens can see the progress. Besides these kitchen table talks, they also organised information evenings where they present their ideas. In cases where farmers have no interest in contacting the representative of the regional water authority, the latter tries to establish the contact through an external person, for example via the LTO (an interest group representing the farmer sector), *natuurmonumenten* or a real-estate broker in the area (Interviews 2017).

When it comes to cooperation, governmental representatives acknowledge that, generally, the individual farmer has little interest in the overall aim of the project. The farmers' focus is mostly on their business and whether certain measures may influence its profitability. According to the respondents, the key skill for an area manager consists in thinking of synergies and how the project may benefit farmers. As one of the respondents pointed out, the area manager needs to be able to find solutions to these problems, s/he needs to have a lot of alternative options in the back of their head, and s/he needs to see synergies between different measures. "You can never say that you can do it. But you can propose to accommodate his/her [the landowners] wishes. […] You have to give people time to think about this" (Interviews 2017).

Coercive Instruments

A **coercive instrument (stick)** would be land ***expropriation***. Many countries have their own set of rules for expropriation for the use of private land for public goals. In general, the Dutch government can oblige private owners to give up their land when there is a clear public interest involved (article 1 Expropriation Act). Examples of such interests can be found in FRM (e.g., in case of dike relocations or creating space for the river like the Hedwige Polder) and also related to projects and plans formalized in land use planning, for example, a new land use designated for creating a housing area or a nature conservation area. The expropriation must be seen as the last resort (*ultimum remedium*) and other remedies must be tried first, such as a fair negotiation on the transfer of the land (amicable acquisition, *minnelijke schikking*, article 17 Expropriation Act).

When government starts negotiating to use private land for public goals, expropriation functions mostly as the big stick in the process. Negotiating a deal is more attractive than expropriation for the landowner; as the first case offers a possibility of exchanging land (land consolidation; *ruilverkaveling*) plus financial compensation while expropriation only leads to financial compensation (Holtslag-Broekhof 2016). The recent investigation of Sanne Holtslag-Broekhof (2016) demonstrates that the expropriation procedure is not a uniform or set case: in many cases the compensation after a court case is higher than the initial offer of the governmental authority involved. Several conditions are connected to the expropriation procedure; for example, expropriation is not legitimate if the landowner can realize the requested provision him- or herself on private land (*zelfrealisatierecht*). Holtslag-Broekhof (2016) concludes that the deterrence that radiates from the Expropriation Act is working well in the

Netherlands, considering the small number of court cases. In most instances, the negotiation leads to a voluntary transfer of land, although she also states that most landowners experience the threat with the big stick as very negative. We might perhaps add that this applies to all parties involved. Most water managers will not be happily involved in a process that might end in an expropriation procedure. According to the interviews, respondents normally tried to avoid confrontations with the landowners and tried to reach out for consensual solutions. This might also be as nature conservation and climate adaptation is not a strong case of public interest.

All in all, expropriation is often very difficult and not the preferred solution for negotiating partners and governmental officials. More differentiated regulations guide expropriation. Firstly, every province has its own provincial regulations that define when land may be expropriated, for example, when 90% of the plots of land are available for a project and only 10% are missing. If the owner is not willing to sell, his/her land may be expropriated. In practice, this is hardly ever applied. Secondly, provinces can expropriate land in the context of Natura 2000. However, within the *Oekense Beek* project, only some smaller plots are Natura 2000 designated areas. If landowners are not willing to change the designation of their land into nature, or do not agree with this change, their land can be expropriated on a relatively short time horizon. Regional water authorities may expropriate land in the context of PAS (*Programma Aanpak Stikstof*) measures. This programme finances measures to reduce nitrogen concentration in the soil. However, there are only two areas where PAS measures may be implemented in this project. The process of expropriation through PAS has to be developed and can take up to four years. However, it appears that in this project PAS expropriation is not viable (Interviews 2017).

Another coercive instrument would be stipulating particular requirements on landowners or leaseholders. The possibilities for governmental authorities to apply this approach are limited. Changing the existing designations of land would be connected to compensation demands. Some farmers also lease land from the municipalities or *natuurmonumenten*. For example, the municipality of Zutphen owns around 40 hectares of land, which is leased to a farmer. The farmer owns extensive livestock and is not particularly interested in alternative measures. The municipality of Zutphen would like to support the project but also has to listen to the interests of their leaseholder (Interviews 2017). *Natuurmonumenten*, in contrast, aims to actively promote sustainable farming. Therefore, they have established particular requirements that leasing farmers need to fulfil, for example, they are prohibited from using pesticides and must comply with any prescribed grazing density (based on e-mail contact with *natuurmonumenten*). Additionally, the farmers have to follow courses on nature conservation management and entrepreneurship ('Natuurbeheer en Ondernemerschap') to receive a certificate by 2021. Here farmers learn how to combine farming activities with nature development and at the same time develop new business models. The aim is that farmers increasingly consider the quality of the soil and the water and take measures to improve it (*Natuurmonumenten* 2017; Kloen 2015). The schooling activities may be considered a communicative instrument (sermons).

Financial Instruments

The main instruments used for dealing with landowners and users are financial incentives (carrots), particularly buying land and/or exchanging land. The province bought small plots of land in the context of the Dutch Nature Network (NNN, Nederlandse Natuur Network). The NNN is a fund provided by the national government that provinces can use to buy land in order to increase the connectivity of ecosystems. The province could already implement measures in these areas. However, as at the moment it is not yet clear which measures are feasible with regard to the water system, they are not yet implementing measures. At the moment, one problem with buying land is the price. The price expectations between the province and the seller do not always align. Another possibility is exchanging land. At the beginning of such a project, the province looked around in the area to find farmers with a lot of land who would like to leave farming, for example, because they have no heir. The province bought some farms and their land. This land could be used for exchange. One provincial interviewee described it as follows: "Some farmers agree to give their land for the project, but they want to have other land in return, land with better soil quality and at a better location, otherwise they would accept money for compensation. In general, the most agreements with farmers can be made based on exchanging land or buying land" (Interviews 2017).

Another financial instrument is compensation. A compensation regulation provides for damages caused by wet conditions; in this case, the province and the regional water authority agreed to share the costs to farmers in the form of lost profit. But this compensation regulation is hardly used. Farmers do not favour this instrument either as it poses too high a risk of uncertainty to their business. However, a representative from the water authority pointed out that it could be combined with flexible solutions, for example, allowing farmers to operate the drainage apparatus. When the farmer wants to use his land, he can operate the drainage, and the wet soil gets drier and workable. At the moment, farmers are not in favour of this approach. One farmer, who is on the administrative board of LTO, thinks about trying this approach on his land as a pilot study (Interviews 2017).

Furthermore, subsidies or payment for particular services offer further financial incentives. In the context of funding programmes such as "*Zoetwater Oost Nederland*", "*Delta Programme Agricultural Waters*" or "*Bodem en ondergrond*" (Soil and Subsoil, a provincial program that aims to improve the soil quality), farmers or private landowners receive payment for cultivating and managing particular vegetation on their lands, such as grassland or alluvial forest. However, as one respondent described, they only receive a subsidy of 70%. Therefore these measures do not always offer a real incentive to a farmer whose fields are cultivated and have no problems with water. Nevertheless, within the project area are also landowners who bought land from the province that was re-designated as nature. They receive money for managing this land. It seems important to give every landowner the same chances. A neighbouring landowner was not offered the possibility to deliver a nature management service in the first round. Now the owner's land is needed for the project, but his/her will to cooperate is limited as s/he was ignored the first time (Interviews 2017).

The municipality of Brummen established cooperation with a local farmer. They want to increase the water infiltration capacity of the soil by increasing the concentration of organic substances. Therefore they use the municipal organic garbage, process it and distribute in the farmer's field. For the municipality, this is more cost-efficient than transporting the garbage across long distances to burn it for biogas production, and the farmer gets paid for this service. As the soil quality is also improved, a representative called it a "win–win–win situation". Besides governmental authorities, *Natuurmonumenten* pays farmers for nature management activities. Some discussion has arisen on how to frame these financial incentives. The municipality of Brummen and other actors talk increasingly about "farmers receiving payment for delivering a service" instead of subsidies or compensation. A representative of the municipality explained that "services describe a clear agreement between actors with rights and responsibilities. It is a real business where taxes have to be paid" (Interviews 2017).

Problem: Support Is Declining

Nowadays stakeholders appear to be less enthusiastic about the project than they were at the beginning in 2013. During information evenings, experts were played against each other and statements were contorted. People may be less enthusiastic for several reasons. Due to the disbandment of DLG, a few years have passed since stakeholders have received any further information about the project. Another issue is mistrust against the government. If governmental officials come to them and talk about PAS measures, they need to mention that expropriation is a possibility. This scares many farmers. Furthermore, some citizens, particularly farmers, feel that the approach of the government is not very consistent. They argue that, decades ago, their family had to give up land to straighten streams, and, presently the government wants land to restore a meandering structure. Related to this, if there are amendments on the plan, one interviewee pointed out that stakeholders pose questions such as "now, what do you actually want? Do you actually know what you are doing?" These issues make the collaboration with private landowners additionally difficult (Interviews 2017).

Reflecting on Challenges of Integrated, Collaborative Projects

The successful implementation of climate change adaptation measures, particularly to address flooding, along the *Oekense Beek* is dependent on the collaboration with private landowners and users. Governmental authorities have several policy instruments at their disposal that may help them find agreements with private landowners.

However, most agreements need to be found on a voluntary base with financial incentives. The governmental actors do not have many coercive instruments at their disposal. "Talking to people" is one of the essentials of this project as one respondent pointed out. Therefore it is necessary to consider the different interests and find compromise. One representative summarized it as follows: "We [citizens] must be able to live comfortably, but we [farmer] must also be able to make money. Compromise and *poldering* [Dutch cooperation approach based on compromise and consensus]."

However, this also leads to a number of challenges or dilemmas for implementing such an integrated and collaborative climate change adaptation project:

- Climate change adaptation is not perceived as an urgent issue by all of the stakeholders. Particularly, in this project, the lack of buffer capacity and the corresponding low groundwater level of the area is actually beneficial for farming activities as the fields are not too wet.
- The long duration of developing such a collaborative project and the iterative adjustment of the project plan due to new stakeholder interests make it difficult to maintain the positive attitude. Stakeholders question the urgency of the project and competence of the governmental authorities, which decreases their own willingness to collaborate. It is also difficult for the governmental authorities, as they need to communicate uncertainties or amendments to the stakeholders.
- The competition of multiple land uses poses additional challenges as the interests and needs of various landowners and users need to be considered. Landowners and leaseholders have rights to cultivate their land. In practice, it turns out to be difficult to provide sufficient financial incentives for farmers to change their approach. At the moment, it appears that another business model or other synergies need to be found in combination with financial incentives. In contrast, nature organisations whose intrinsic interests align with the aim of the project are more inclined to implement measures. One may even speculate that they would have done so anyway if they had the financial means.
- The development of innovative synergies is not always supported and sometimes downright hindered by the existing institutional structures. For example, the use of organic garbage to raise the concentration of organic substances in the soil is difficult to implement, as the existing waste management regulations do not consider these synergies. The flexible exchange of land is also sometimes hindered by strict regulations.
- The establishment of voluntary agreements with private landowners is sometimes negatively influenced by the legal need to inform. For example, when governmental authorities approach landowners in the context of the PAS measures, they have to disclose that, legally, expropriation is a possibility even though the regulations may not apply to this specific project. This scares off the landowner and negatively influences his/her willingness to collaborate in the project.
- It further seems important to treat the various landowners equally; otherwise they may be uncooperative in the subsequent process of the project, as they feel mistreated. However, this may be difficult as landowners have divergent interests, which sometimes ask for tailor-made solutions.

Acknowledgements Open access of this chapter is funded by COST Action No. CA16209 Natural flood retention on private land, LAND4FLOOD (www.land4flood.eu), supported by COST (European Cooperation in Science and Technology).

References

Ansell C, Gash A (2007) Collaborative governance in theory and practice. J Public Adm Res Theor 18(4):543–571

Arts B, Leroy P, Arts B (2006) Institutional dynamics in environmental governance. Springer, Dordrecht

Baird J, Plummer R, Bodin Ö (2016) Collaborative governance for climate change adaptation in Canada: experimenting with adaptive co-management. Reg Environ Change 16(3):747–758

Bemelmans-Videc ML, Rist RC, Vedung E (2010) Carrots, sticks and sermons: policy instruments and their evaluation. Comparative policy evaluation series, p 277

Biesbroek GR, Dupuis J, Jordan A, Wellstead AM, Howlett M, Cairney P, Rayner J, Davidson D (2015) Opening up the black box of adaptation decision-making. Nat Clim Change 5(6):493–494

Collins K, Ison R (2009) Jumping off Arnstein's ladder: social learning as a new policy paradigm for climate change adaptation. Environ Policy Gov 19(6):358–373

Driessen PPJ, Dieperink C, Van Laerhoven F, Runhaar HAC, Vermeulen WJV (2012) Towards a conceptual framework for the study of shifts in modes of environmental governance—experiences from the Netherlands. Environ Policy 160:143–160

Duit A (2015) Resilience thinking: lessons for public administration. Public Adm 94(2):364–380

Holtslag-Broekhof S (2016) Private grond voor Publieke doelen. Uitgeverij BOXPress

Kaufmann M, Van Doorn-Hoekveld W, Gilissen HK, Van Rijswick M (2016) Drowning in safety—analysing and evaluating flood risk governance in the Netherlands. STARFLOOD Consortium, Utrecht

Kloen H (2015) Van Pacht naar Partnerschap. Landwerk 3:5–9

Brummen Landschapsnetwerk (2013) Advies bij DLG-studie klimaatcorridor en verbindingszone Oekense beek, Brummen. LNB, Brummen

Natuurmonumenten (2017) Gemeenschappelijk landbouwbeleid. Available via DIALOG. https://www.natuurmonumenten.nl/standpunten/gemeenschappelijk-landbouwbeleid. Accessed 18 Dec 2017

Olsson P, Gunderson LH, Carpenter SR, Ryan P, Lebel L, Folke C, Holling CS (2006) Shooting the rapids: navigating transitions to adaptive governance of social-ecological systems. Ecol Soc 11(1)

Gelderland Provincie (2012) Water shortage and climate adaptation in the Rhine Basin. Provincie Gelderland, Arnhem

Gelderland Provincie (2016) Beheerplan Natura 2000 58—Landgoederen Brummen. Provincie Gelderland, Arnhem

Sjöstedt M (2015) Resilience revisited: taking institutional theory seriously. Ecol Soc 20(4)

Stroming (2013) A green Rhine corridor. Future proofing Western Europe's largest river for people, nature and the economy

Termeer CJAM, Dewulf A, Breeman G (2013) Climate change governance. In: Knieling J, Leal Filho W (eds) Climate change governance. Springer, Berlin, Heidelberg, pp 27–39

Wise RM, Fazey I, Stafford Smith M, Park SE, Eakin HC, Archer Van Garderen ERM, Campbell B (2014) Reconceptualising adaptation to climate change as part of pathways of change and response. Glob Environ Change 28:325–336

Maria Kaufmann is an Assistant Professor in Environmental Governance at the Radboud University in Nijmegen, The Netherlands. Maria has a multidisciplinary background holding an MSc degree in Environmental Sciences and a Ph.D. degree in Management Sciences. Maria's research focuses on examining environment-society interactions in the context of climate adaptation, particularly flood risk governance. She is interested in analyzing institutional dynamics and evaluating governance arrangements in terms of environmental justice.

Mark Wiering is Associate-Professor in Environmental Governance and Politics at Radboud University in Nijmegen, The Netherlands. He is interested in societal transformations in light of sustainability and the role environmental governance therein. His domains are, more specifically, the dynamics of governance in FRM and water quality management and the dynamics related to grass roots movements in renewable energy.

Chapter 22
Commentary: Effectiveness and Integrated Multi-use of Retention Measures—A Hydraulic Engineering Perspective

Robert Jüpner

As a hydraulic engineer with a research focus on FRM, I read the chapter "Dilemmas of an Integrated Multi-use Climate Adaptation Project in the Netherlands: the Oekense Beek" with great interest. The described case in the Netherlands seems to me a typical river restoration project that aims to fulfill the requirements of the European Water Framework Directive. Since the implementation of this directive, a number of Europe's rivers have been modified to improve their ecological status. Most of the projects focused on hydromorphological and biological quality elements. Updated case studies are published for a broader community, for example at https://restorerivers.eu.

The authors of the article provide few technical details about the size and the engineering part of the project. To fully understand the described "multi-use climate change adaptation project" from a hydraulic engineering perspective, the following information is essential:

- size of the project area, especially the relevant hydraulic and hydrological parameters,
- calculated retention volume for various flood scenarios,
- quantification of the described effects after successful implementation.

Even if there is no clear "threshold", for a restoration project to be considered effective, the retention volume needs to be significant in proportion to the flood discharge. In engineering practice, hydraulic models will be used to calculate the reduction of the flood peak as the most important effect of increased retention capacity. The detailed figures thereby depend on a number of parameters, for example, the shape of the flood hydrograph and the filling procedure of the retention area itself (Patt and Jüpner 2013).

R. Jüpner (✉)
Hydraulic Engineering and Water Management, Technische Universitaet Kaiserslautern, Kaiserslautern, Germany
e-mail: Robert.Juepner@bauing.uni-kl.de

T. Hartmann et al. (eds.), *Nature-Based Flood Risk Management on Private Land*,
https://doi.org/10.1007/978-3-030-23842-1_22

Nevertheless, the core aim of the Oekense Beek project is clearly addressed: the restoration of a straightened and deepened stream and part of its floodplain. This will lead to an improvement of some parameters, for example, the quality of aquatic habitats, the variety of ecomorphologic parameters, more diverse flow conditions, and a rise in the ground water level. In addition, the retention capacity of the area will increase. However, due to the size of the project, these improvements will most likely be limited to the project region itself.

The effect of small- and medium-scale retention measures is subject of numerous research projects (Burek et al. 2012; Collentine and Futter 2016; Reinhardt et al. 2010; Rieger and Disse 2008). Even if there is no doubt about the general benefit of nature-based flood risk measures, the size of the flood peak reduction is usually very small. Meanwhile, multicriterial modelling of the effectiveness of decentralized flood protection measures are available (Neumeyer et al. 2018). The results are explicit: The more flood water needs to be considered the more limited the effect will be.

The illustrated Oekense Beek project is described as a "multifunctional-use-project". While this constitutes progress, one must ask what are the main goals? A clear distinction between primary effects, such as the improvement of habitat for some species and positive "side effects", will help to accurately measure the success of the project.

The effect of small- and medium-scale restoration projects toward reducing the negative effects of floods and droughts will diminish as the size of flood events increase. Various authors describe this phenomenon in detail (Burek et al. 2012; Patt and Jüpner 2013). Nevertheless, a positive result regarding FRM will certainly be achieved if the project is implemented successfully, because the overall "buffer capacity" will rise in the project area.

The statement that the "increase of precipitation [due to climate changes] can be compensated" is a very positive point of view. Most likely, a precise quantification of these effects will be very difficult (DWA 2015; Burek et al. 2012). Again, the more water needs to be retained during a rare flood event, the more limited the effects of (small-scale) nature-based flood retention measures are. In DWA (2015), quantitative results for the effectiveness of different nature-based flood retention measures are published. The results are related to the reduction of a flood event with a hundred-year recurrence interval. Based on 20 practical examples, the effectiveness of river restoration measures ranges from 0 to 30% with an average of 8.5% (DWA 2015, p. 79). Nevertheless, a very positive effect on the more frequent small flood events is verified.

In the case description, the cooperation between the relevant governmental and non-governmental actors is described in detail. This very interesting analysis is of great importance, because usually engineers are not specifically informed about this background of a project, whereas landscape and urban planners are more familiar with the project's general background. Understanding the motivation and main ideas of all actors/stakeholders will facilitate developing a project that can be implemented successfully. Usually, a variety of engineering approaches are available to reach the main project goals. The more effort spent in the early stages, the smoother the realization of the project.

In the case of the Ookense Beek, conflicts emerged between various actors within the project. This seems typical and representative for river restoration projects. It is very interesting to read, which (legal) approaches are available in the Netherlands and how they are used. The explained effectiveness of the tools over time gives additional helpful information. This is especially significant, since the legal framework—the European Water Framework Directive—is the legal basis in all EU-member states, where "core values" are mostly identical. It is extremely beneficial for hydraulic engineers to learn from other European States how to efficiently use these instruments to reach project goals.

Acknowledgements Open access of this chapter is funded by COST Action No. CA16209 Natural flood retention on private land, LAND4FLOOD (www.land4flood.eu), supported by COST (European Cooperation in Science and Technology).

References

Burek et al (2012) Evaluation of the effectiveness of natural water retention measures. In: Burek P, Mubareka S, Rojas R, de Roo A, Bianchi A, Baranzelli C, Lavalle C, Vandecasteele I (eds) JRC scientific and policy reports. Publications Office of the European Union, Luxembourg. https://doi.org/10.2788/5528

Collentine D, Futter MN (2016) Realising the potential of natural water retention measures in catchment flood management. In: Collentine D, Futter MN (eds) Trade-offs and matching interests. J Flood Risk Manag. https://doi.org/10.1111/jfr3.12269

DWA (2015) DWA-Merkblatt 550 "Dezentrale Maßnahmen zur Hochwasserminderung". Deutsche Vereinigung für Wasserwirtschaft, Abwasser und Abfall. Hennef. ISBN 978-3-88721-262-9

Patt H, Jüpner R (eds) (2013) Hochwasser-Handbuch; 2., neu bearbeitete Aufl. Springer, Vieweg. ISBN 978-3-642-28190-7 (in German)

Reinhardt C, Bölscher J, Schulte A, Wenzel R (2010) Decentralised water retention along the river channels in a mesoscale catchment in south-eastern Germany. Phys Chem Earth 36(2011):309–318

Rieger W, Disse M (eds) (2008) Wasserrückhalt in der Fläche – Möglichkeiten und Grenzen des dezentralen Hochwasserschutzes. Mitteilungsheft des Instituts für Wasserwesen, Universität der Bundeswehr München. Heft 100/2008, ISBN 978-3-8356-3173-1. (in German)

Neumeyer M, Heinrich R, Rieger W, Disse M (eds) (2018) Vergleich unterschiedlicher Methoden zur Modellierung von Renaturierungs-und Auengestaltungsmaßnahmen mit zweidimensionalen hydrodynamisch-numerischen Modellen. Conference Paper, Tag der Hydrologie 2018, Dresden. Available via DIALOG. https://www.researchgate.net/publication/325415059_Vergleich_Unterschiedlicher_Methoden_zur_Modellierung_von_Renaturierungs-_und_Auengestaltungsmassnahmen_mit_Zweidimensionalen_Hydrodynamisch-Numerischen_Modellen. Accessed 6 Nov 2018

Robert Jüpner studied Hydraulic Engineering at Technische Universität Dresden and got a Ph.D. from Freie Universität Berlin, Germany. He has an appointment as a Professor for Hydraulic Engineering and Water Management at the Department of Civil Engineering of Technische Universität Kaiserslautern, Germany. For more than 20 years, he has worked in the field of FRM and flood protection and has gained specific expertise during his work in emergency management of big flood events in Germany.

Chapter 23
Commentary: Voluntary Agreement in Multi-use Climate Adaptation in the *Oekense Beek* from a Politic-Economic Perspective

Thomas Thaler

Introduction

Recent high impact floods and droughts were experienced across the EU, where the economic and social impact was significant (Guha-Sapir et al. 2016): more than five times the losses incurred between 2000–2012. Driven by climate change, extreme flooding events, such as those experienced across the EU, are expected to increase in frequency with models suggesting that average annual economic losses predicted to exceed EUR 23 billion over the same period (Jongman et al. 2014). Preparing for and building resilience against future natural hazards events is challenging and resource-intensive (e.g., time, financial, etc.) with key difficulties for practical application. This era of climate change calls for new robust (i.e., inclusive of known or probable risks) and flexible (i.e., incorporating uncertain or possible risks) risk management approaches. The desire to manage land and water sustainably, introduce resilience to climate change, assess risk and implement sustainable environmental management strategies has broad support, but defining "sustainable" management has proven difficult for policymakers. One adaptation strategy might be Nature-based Solutions (NBS). NBS aim to harness ecosystems through both their resilient adaptive eco-services and sustainable integrity that will provide short-, medium- and longer-term solutions to managing the risks associated with climate-driven extreme hydrometeorological events being experienced across Europe and globally. Nature-based solutions for hydrometeorological risk mitigation and adaptation in river catchments involve, for example, Natural Water Retention Measures (NWRM), space for the rivers, or measures for resilient cities (i.e., green infrastructure in cities, green roofs, decentralized rainwater management). These solutions are also referred to as "green and blue infrastructure". Nature-based solutions to water-related risks cannot entirely

T. Thaler (✉)
Institute of Mountain Risk Engineering (IAN), University of Natural Resources and Life Sciences (BOKU), Vienna, Austria
e-mail: thomas.thaler@boku.ac.at

T. Hartmann et al. (eds.), *Nature-Based Flood Risk Management on Private Land*,
https://doi.org/10.1007/978-3-030-23842-1_23

substitute for traditional measures such as flood pathway and receptor approaches both structural and behavioral (e.g., flood walls, channels, flood warnings), but their potential value for mitigation and adaptation has been recognized (Lafortezza et al. 2018). As such, NBS can often be easily designed in engineering terms and provide a good complement to local climate adaptation strategy, but a limiting factor is the area of land required to provide sufficient storage in the appropriate place to be useful. Nevertheless, implementing and paying for NBS and for grey or mixed (grey and green) infrastructure, requires not only appropriate technical data, risk analysis, functional testing but requires funding for building and maintaining the various targeted options over variable time-space continuums. There are significant differences between countries worldwide on how NBS must be implemented; but generally at least three main barriers can be summarized:

- Cultural and social barriers: for example, in England and Wales FRM policy, the political definition has not so far been publically acceptable to use private land (upstream) to sacrifice it for the downstream communities as a mainstream strategy (Thaler 2015).
- Uncertainties in frequency and magnitude: using flood storages as key FRM scheme also causes significant uncertainties about the next event, which cause large concerns by private landowners about how to use the land.
- Mechanisms of compensation: flood storage includes the challenge of transferring a risk and benefit to others. This causes complicated discourse about the preferred form and institutional set up of compensation. For example, Ungvari and Kis (2013) showed that the implementation of flood storages on the Tisze river basin is that farmers and government have different views on how to organise the payment scheme (small fixed annual amounts or large amount based on the event).

Use of Economic-Policy Instruments in Flood Risk Management

As a result, the change in ownership such as land buying or taking by land expropriation might be approached to implement NBS. One solution might be to use/implement Economic Policy Instruments (EPIs) to manage easier water-related risks. EPIs have become more popular in the past decade, in particular, with the implementation of various EU regulations and directives. The range of EPIs can be as follows: (1) innovative payments schemes (i.e., compensation by public administration or insurance); (2) financial incentives for land-use changes (e.g., agro-environmental schemes); (3) flood risk pooling schemes; (4) financing schemes for urban development for stormwater management; (5) voluntary agreements, for example, between urban and rural areas; and (6) cap and trade schemes, like the insurance bonus malus system (Thaler 2015). In the Oekense Beek study area, the aim was to use the EPI voluntary agreement between private landowners (i.e., farmers) and users, such as regional water authority, the province, municipality Brummen as well as the

Table 23.1 Framework of analysis: revealing opportunities in multi-use climate adaptation

	Focus of research questions	Short description	Revealing opportunities
Natural capital	Focus on flood event (hydrology science)	• Catchment characteristics • Flood characteristics • Retention volume capacity	Landform engineering
Social capital	Focus on stakeholders (sociology science)	• Flood risk perception • Solidarity and trust • Social network	Participation process
Institutional framework	Focus on power (political science)	• Formal/informal rules • Administrative boundaries	Pilot removes such barriers
Socio-economic activities	Focus on funding (economic science)	• Socio-economic losses	Compensation schemes

natuurmonumenten. First of all, the definition of a voluntary agreement is open to argument (OECD 2003) and is often simply defined as any approach that does not involve a legally enforceable requirement that is imposed on one party to take action in the interests of another. Nevertheless, capturing, analyzing and understanding differences of success and failure in the adoption of voluntary agreement is challenging (Thaler 2015) yet necessary to encourage and support stakeholders' engagement in this direction and to understand limits in the transferability of success from one case to another. The successful implementation of voluntary agreements is clearly influenced (positively or negatively) by a number of factors. These factors may constrain the feasibility and acceptability of a project but also determine its efficiency. Providing and pre-empting an exhaustive list of factors would be misguiding. However, these factors can be grouped into four categories: (a) natural capital, (b) social capital, (c) institutional framework and (d) socio-economic activities (see Table 23.1).

Natural capital refers to the stock of resources that provides environmental services. The catchment characteristics play a central role to implement flood storage areas. The retention volume capacity for instance will depend on various natural factors such the landscape profile, the soils and the geological conditions (EA 2009). The volume capacity also needs to match the hydrological behaviour of the catchment, the considered river and the upstream and downstream tributaries. A central aim of using catchment-wide NBS is to protect high value/vulnerability (usually urban) areas in the lower part of the catchment. Therefore, the central question and conflict arise how to compensate low-intensity agricultural areas and how to motivate farmers to provide the land as in the Oekense Beek example. Behind this simple transfer—interesting from a flood management and an economic perspective—lie potential sources of social tension and resistance to the project as the transfer from one location to another (e.g., often from one community or administrative boundary to another) means risk. The pre-existence or prior lack of shared knowledge, trust and social connection may influence the acceptability of such transfer. A critical element is the interaction (*social capital*) between farmers and other stakeholders from urban areas. Most of the time, the examples show a lack of social capital of these two groups by a lack of risk culture and/or solidarity. One recurrent conflict is the impact that adopting environmentally friendly measures may have on the *socio-economic activities* for farmers, as it reduces the profitability of a business by internalising externality. There is the question how to organise the compensation, which is often based on a negotiation process. However, such negotiations are regulated by *institutional framework* (Ostrom 1986; Scott 1995), formal and informal, which may affect the implementation of a policy instrument. An understanding of the possibility and limits within the institutional framework is crucial. In the context of implementing flood storage areas, it is essential to investigate the question of property right (e.g., right to flood/right to be protected), to land use planning (e.g., flood-prone areas defined or not) and to existing policy on the funding mechanism (e.g., right to compensate). Voluntary agreements involving a form of compensation are often preferred; yet their implementation differs from one place to another (in particular challenging the upscaling or transfer the lessons learn to other cases).

In seeking to construct a flood storage area, there are alternative forms of power that might be used, with associated advantages and disadvantages. Because flood storage requires space and place, a purely market-based approach cannot always be relied upon to assemble the necessary area required in the appropriate place. In addition, the rules covering action by a particular administrative unit commonly will not allow it to buy land outside of its administrative boundaries. This is usually even more the case when the expropriation of land is concerned; for example, the Netherlands does not have the legal powers to acquire land in Germany through compulsion in order to construct a flood storage area. In strongly federal states, the same is true between federal states. Hence, a voluntary agreement may be the only viable option when action is desired outside of the boundaries of proposing body.

Alternatively, if an attempt to use power is likely to be met by resisting power, there are two reasons why a voluntary approach may be preferable. Firstly, even if the resistance can be overcome, this will involve costs and time delays. Secondly, creating an adversarial context in one case can incur long-term costs by creating the anticipation by one or several parties that the future will also be one of conflict even when the interests of the parties actually coincide. Conversely, establishing cooperation in one instance may create a precedent for future cooperation and a norm of reciprocity (Nowak and Highfield 2011).

Conclusion

In FRM, voluntary agreements are now also receiving a lot of interest as complements to the existing policy instruments in order to achieve the objectives the EU

Water Framework Directive and of the EU Floods Directive, such as the implementation of flood storages and use of natural retention areas. Whilst the issues of scale and fit of administrative units to physical problems have been identified (Underdal and Young 1997), if it is impractical or inappropriate to change the boundaries of the administrative, it is necessary to create bridging mechanisms (Kohn 2008) to enable co-action across the boundaries of the existing administrative units. Thus, the central problems addressed in the Oekense Beek and many other examples are the appropriate use of power and bridging across the boundaries to power created by rules. At the same time, the use of economic instruments has been questioned from the perspective of social equity (Thaler and Hartmann 2016). The importance of equity and distributional issues (be it between water use sectors, social groups or regions) is in fact receiving increasing attention in many policy discussions and research activities in different parts of the world. Not all these objectives are fulfilled to the same extent by the different economic instruments. More often than not, in practical policy implementation more attention is paid to use of EPI as a mean to raise revenues than to efficient allocation of water use/water service delivery.

Acknowledgements Open access of this chapter is funded by COST Action No. CA16209 Natural flood retention on private land, LAND4FLOOD (www.land4flood.eu), supported by COST (European Cooperation in Science and Technology).

References

EA (2009) Achieving more, operational flood storage areas and biodiversity—final report. Environment Agency, Bristol

Guha-Sapir D, Hoyois P, Wallemacq P, Below R (2016) Annual disaster statistical review 2016. The numbers and trends. Université catholique de Louvain, Brussels

Jongman B, Hochrainer-Stigler S, Feyen L, Aerts JCJH, Mechler R, Botzen WJW, Bouwer LM, Pflug G, Rojas R, Ward PJ (2014) Increasing stress on disaster-risk finance due to large floods. Nat Clim Chang 4:264–268

Kohn M (2008) Trust: self-interest and the common good. Oxford University Press, Oxford

Lafortezza R, Chen J, Konijnendijk van den Bosch C, Randrup TB (2018) Nature-based solutions for resilient landscapes and cities. Environ Res 165:431–441

Nowak M, Highfield R (2011) Super cooperators. Canongate, Edinburgh

OECD (2003) Voluntary approaches for environmental policy: effectiveness, efficiency and usage in policy mixes. OCED, Paris

Ostrom E (1986) An agenda for the study of institutions. Public Choice 48(1):3–25

Scott RW (1995) Institutions and organisations. Sage, London

Thaler T (2015) Rescaling in flood risk governance—new spatial and institutional arrangements and structures. Middlesex University, London

Thaler T, Hartmann T (2016) Justice and flood risk management: reflecting on different approaches to distribute and allocate flood risk management in Europe. Nat Hazards 83(1):129–147

Underdal A, Young OR (1997) Institutional dimensions of global change. IHDP Scoping Report

Ungvari G, Kis A (2013) Floods and water logging in the Tisza river basin (Hungary). WP 4 Ex-Ante Case Studies. EPI-Water, Venice

Dr. Thomas Thaler is a Research Fellow at the Institute of Mountain Risk Engineering, University of Natural Resources and Life Sciences. His research focuses on the topic of politico-economics and natural hazards in Europe, with a particular emphasis on questions relating to design and effectiveness of governance systems as well as integrating European environmental policies into national and local institutions.

Part IX
Conclusion

Chapter 24
Conclusion

Thomas Hartmann, Lenka Slavíková and Simon McCarthy

Private land matters in FRM. In particular, private land is closely associated with NBS in FRM—*nature-based flood risk management*. Nature-based solutions are currently receiving a large degree of attention in policy, academia and slowly in practice (see introduction). These measures need more land, and this land is often privately owned. However, experience of implementing NBS in FRM remains scarce; this book showcases much called for empirical practice examples of nature-based FRM on private land.

The examples from different parts of Europe illustrate the wide variety of NBS that are currently available, but they also show the variety of private land issues that can arise on various scales. Looking at the examples shows us that privately owned land in FRM does not necessarily mean that the land is legally owned by an individual person; also public authorities can be owners under a private law regime. Within this volume, when private land is referred to, the term land refers to that which falls under private law. Private law regulates interactions of legal persons (on land)—opposed to public law, that applies to the relation between public authorities and private legal persons (Needham et al. 2018). This means, individuals, regional

T. Hartmann (✉)
Faculty of Environmental Sciences, Wageningen University & Research, Wageningen, The Netherlands
e-mail: thomas.hartmann@wur.nl

Faculty of Social and Economic Studies, J. E. Purkyně University, Ústí nad Labem, Czech Republic

L. Slavíková
Faculty of Social and Economic Studies, Institute for Economic and Environmental Policy (IEEP), J. E. Purkyně University, Ústí nad Labem, Czech Republic
e-mail: lenka.slavikova@ujep.cz

S. McCarthy
School of Science and Technology, Flood Hazard Research Centre, Middlesex University, London, UK
e-mail: S.McCarthy@mdx.ac.uk

T. Hartmann et al. (eds.), *Nature-Based Flood Risk Management on Private Land*,
https://doi.org/10.1007/978-3-030-23842-1_24

and local self-administrative units and national/federal states might all serve in the role of landowners pursuing different interests. So, ultimately the book addresses numerous challenges to implementing NBS with the focus on different ownership and planning structures, scales and contexts across Europe.

It becomes evident from this volume, with commentary insights from different disciplinary perspectives, that more land is needed for the implementation of such measures. This raises implications for multidisciplinary research, transdisciplinary knowledge coordination and more intensive stakeholder engagement. Fragmented knowledge and practice domains characterized by a wide range of disciplines (in land-use planning, hydrology, property rights, economics, sociology, ecology, landscape planning, policy science to name a few) are required to not only plan for technically viable approaches, but also to gain social consensus to provide the access to the necessary privately owned land. How such land is acquired may also have implications in long-term cooperation between stakeholders for the sustainable maintenance and further adoption of NBS. Land access and cooperation necessitates relationships between the various concerned stakeholders. It is revealed how the stakeholder interrelationships are important. These can be driven or frustrated by formal policy, legal and economic instruments and by levels and types of knowledge and experience. In addition, informal activities to engage landowners and related decision-makers appear to influence and play an important role in implementation as initially these parties have little or no experience with implementing nature-based FRM on private land.

The examples presented in the book draw mainly on experiences across North-West and Central Europe addressing differing contextual and implementation approaches within a range of topographies and scales. This volume includes examples of nature-based FRM from Austria, Belgium, Czech Republic, Germany, Netherlands and Poland. Scales vary from small and local retention measures with narrow ownership structures (see for example the Czech case with only one landowner involved) up to the catchment level, where planning and wider stakeholder engagement challenges implementation (see the cases from Austria or Germany). All the cases express specific local complexities and are highly contextual. However, broad questions can be identified that cut across the cases supported by the reflections of the expert commentators on each case. The selection of cases was based on the idea of covering a huge bandwidth of NBS in different contexts (without eliminating these contexts). General conclusions from the different disciplines indicate that some of the issues are cross-cutting and more related to disciplinary rather than country-specific issues.

So what are the broad cross-cutting issues that have been identified? Well, namely how ownership matters in nature-based FRM, how processes for implementing such measures need to be facilitated, the aspects of time and scale, but most importantly the communication across disciplines.

Ownership of Land Matters

The examples illustrate that the amount of additional land needed for FRM is substantial when using NBS. Implementation requires engaging with and gaining the support of the landowners. The cases prove that nature-based FRM is land-intensive and that the legal aspects of the land ownership (i.e., full property rights, tenure, and other sticks in a bundle of rights), the number and type of owners (i.e., public authorities or private persons) matter for successful implementation. The reconciliation of public and private interests supported with the appropriate planning tools and funding strategies is crucial—as the Belgian case illustrates particularly, but also Mr. Pitek's land.

In cases where public authorities or the state are owners of the land, as in the small retention programs in the Polish forests, implementation and possible upscaling tends to be smoother. As the planning and funding authority overlaps with the land ownership. But this relies on those public authorities taking the lead and, as in this case, being motivated by wider political or market conditions (see Futter, this volume) or more usually through economic justification in planning to persuade state support (see Macháč, Louda and Löschner, this volume). However, even without the financial support of the state, nature-based FRM is feasible, as the case of Mr. Pitek's land illustrates. The self-motivated private landowner created retention ponds on his private land (see Slavíková and Raška, this volume). A private landowner challenging public policy by undertaking what "he feels is the right thing" can be considered an unusual case (see Löschner's commentary this volume) counter to the more common view taken by landowners: "why would I want to pay for that?" (usually in the context of top-down implementation) (see Kapović Solomun, this volume). Both the latter commentaries highlight how important it is that landowners and users, even when highly motivated, feel engaged in the decision making. It is considered vital for the implementation of nature-based FRM that it does not run counter to wider flood management strategy, or, as in the example provided of a Scottish landowner (see Wilkinson, this volume), may run against community practices. The landowner might be self-motivated to pursue NBS but more often requires persuasion and sometimes formal agreements to facilitate their involvement. So, we can conclude from the experiences described in the cases, and reflected upon in their respective commentaries, that land does indeed matter for nature-based FRM, but this does not automatically entail that authorities responsible for FRM do always need to *own* that land.

Facilitating Nature-Based Solutions

Following on from this remark on the ownership of land for nature-based FRM, the question is raised of what we can learn from the examples in terms of facilitating implementation. It is apparent in the chapters that facilitating nature-based FRM requires the engagement and agreement of landowners for respective changes in land use. Ultimately, it is the individual landowner who needs to accept interventions in land uses or even implement the changes themselves. This asks for strategic planning at a regional and catchment level to take land use change from the landowner's point of view into consideration. The examples from the different countries and at different scales reveal that barriers to implementing measures can stem from uncertainties in hydrological effectiveness and mechanisms of compensation, but that also cultural and social aspects matter (see Thaler, this volume). All these issues need to be addressed as well as landownership. Landowners, and often decision-makers (such as mayors or local authorities), may have little or no experience in the facilitating process for nature-based approaches (van den Brink 2009). Both in the case of turning private agricultural land back into floodplain forests in Germany (see Warner, this volume) and in response to climate adaptation as in the Netherlands case (see Kaufmann, this volume), authors explore the challenges of gaining stakeholder acceptance via a range of engagement practices and financial incentives. Such engagement practices are broader than the traditional cost-benefit arguments and embrace methods of stakeholder involvement and public engagement. This essentially means embedding methods and techniques from different disciplines such as legal governance, planning, social science and economics, in working with landowners and other stakeholders related to the land needed. It is clear from the cases that one set of engagement approaches will not fit all contexts but an appreciation that a variety of methods will need to be employed tailored to the stakeholder context that presents itself. In the introduction, we stated that a key question for implementation is the justification of NBS and for that discussion to take place, empirical evidence is required and a wider understanding of the constraints and enablers beyond just the hydrological impacts is essential.

Time and Scale Matter

So, as indicated earlier, different stakeholder and often related professional disciplines have different perspectives on time and scale and thus regarding the relationship of NBS and their interaction with land. Scale is a crucial aspect of these perspectives but is clearly interrelated with the temporal. The cases reveal differing scalar issues and some of the challenges with such multiple scalar perspectives on nature-based FRM.

There are certainly limitations to the ability of how nature-based FRM methods contribute on their own to the scale of flood risk reduction; their strength lies in the mitigation of more frequent lower inundation events (see Futter, this volume). Even the dyke relocation in the Elbe-Brandenburg case (see Warner and Damm, this volume) is contested (see Staveren, this volume). The same commentator highlights the spatial and temporal characteristics that differentiate fluvial and coastal geographies in terms of water type, ecosystems and seasonality influences, both hydrological and ecological characteristics. At the same time, the Elbe-Brandenburg case illustrates how the more immediate issue of flood prevention is a stronger aim than nature restoration or loss of land. In other cases, engineering perspectives focus on the underlying argument of understanding and quantifying the reduction in the flood hydrograph and the limitations of smaller-scale nature-based applications. In a nutshell, the more immediate water is needed to be retained in a flood event, the less effective NBS are (see Jüpner, this volume). This hydro-engineering perspective strongly challenges the "common sense" view that every retained raindrop counts and perhaps introduces the temporal aspect of where in the hazard cycle and in relation to the event retention takes place. However, nature-based FRM can also go beyond slowing the flow of flooding (see Kapović Solomun, this volume). Rather than focus just on the FRM aspects, additional longer-term and even perhaps larger-scale ecological and environmental benefits can be brought to the fore but need to be clearly defined. Regulation of droughts and the overall benefit of improving the "buffer capacity" of environmental project areas can be considered (see Jüpner, this volume). Maintaining water quality may be considered but understanding the geology for groundwater recharge is required (see Wilkinson, this volume). It is clear that wider benefits can be multifunctional—including biodiversity, recreation and water management at the local scale but also globally via the hydrological cycle (see Futter, this volume). For those effects to be substantial, scale matters for nature-based FRM but implementation and contribution in that management may be and in some cases is currently driven more by these other benefits. This later point will be returned to later in this chapter.

This volume illustrates that nature-based FRM necessitates multi-disciplinary and cross-disciplinary cooperation not only in the physical and social sciences, but also in the legal and policy/planning arenas. However, institutional and project management may frustrate such interaction. In terms of the scale of the issue interestingly, the observation that, as engineers, they are rarely exposed to the same stakeholder information as landscape and urban planners reveals the potential inequalities in access to information among the project's professional stakeholders (see Jüpner, this volume). It also becomes apparent that different disciplinary perspectives work to differing time frames. The functionality and impact of NBS can take more time to develop than that of the conventional hard engineering schemes. A sound way to evaluate these specific temporal and scalar challenges of nature-based FRM may currently not be accounted for in contemporary planning instruments in FRM (see Löschner et al. this volume).

The cases do not represent a comprehensive overview of all possible types of measures, but they show the tensions of nature-based FRM at various scales and in different time frames compared to traditional flood protection measures. So, one lesson that can be drawn from these cases is that of the roles of time and scale and their different requirements across the wide range of stakeholders impacts implementation.

Communication Across Disciplines Matter

The majority of authors in this volume continue the call for better empirical information on the impacts and effectiveness of nature-based FRM to support the consensus building and subsequent implementation. There is still a lack of proof regarding the degree of reduction in the hydrograph for varying return periods combined with the additional benefits (restoring urban wetlands) (see Pohl, this volume). Ex-post monitoring of projects for effectiveness would be invaluable to inform such decisions (see Veidemane, this volume). Decisions remain supported by traditional cost-benefit approaches of which one is explored in the urban wetlands in the city of Pilsen in the Czech Republic. The possibly greater contribution comes in terms of social benefits set against costs from NBS is revealed (see Macháč and Louda, this volume). However, Gutman (this volume) highlights a lack of research on the *perceived* effectiveness and legitimacy of the implementation of nature-based FRM. Currently nature-based FRM might be viewed as approaches for the restoration of multiple ecosystem services rather than measures for flood risk mitigation. This perhaps reveals how the political context and institutional agenda can drive different strategies undertaken in European countries. These insights are particularly valuable for engineering and hydrology, fields that still tend to underestimate the role of social constructionism and respective multiple perceptions on actual implementation of measures and their effects. It can be said that disciplines of engineering and hydrology might need to learn to communicate differently with landowners and other "non-expert" persons in FRM. This also became clear in the preparation of the volume, when first drafts of commentaries from engineering or hydrological disciplines were considered rather short and technical by the editors (different disciplinary backgrounds). This communication aspect (not possible for the commentators in isolation) is especially relevant as these disciplines naturally play a key (but shared) role in the justification and implementation of nature-based FRM decided by stakeholders with multiple interests.

Also for economic and legal aspects, it is essential that instruments and effects need to be communicated well. This can be illustrated with the challenges of swapping development rights in the Flemish case where the high transaction resource requirements of this approach meant that the original zoning approaches in place were eventually the preferred option (see Crabbé and Coppens, this volume). This case highlights how economic instruments can be undermined if a gradient does not already exist in the market from supply to demand. Here this does not appear to be the case, and market interventions are proposed (see Kis and Ungvári, this volume) revealing how the contextual challenges for such instruments require as much attention as the measures themselves. However, market intervention may not always be palatable to decision-makers. In this case, there appears to be a lack of willingness to engage with the market rather than government-led initiatives assuming a distinction between public and private property rights that appears unsurmountable (see Sheehan, this volume). Generally, where the legal opportunities are present, the economic policy instruments provide a less disruptive approach in terms of financial, temporal and sustainable stakeholder relationships (see Thaler, this volume). Still, it is imperative that such instruments are communicated well and in a way that "non-experts" can comprehend the consequences and implications.

The same applies for other disciplines involved and their efforts to communicate. Some disciplines or individuals through the necessity of their work appear to have recognized the benefits of cross-disciplinary communicating more than others (i.e., spatial planning, geography or social science). The need to involve multiple disciplines in nature-based FRM on private land essentially requires an appreciation and use of a commonly understood language or at least a way of interdisciplinary communication that takes other stakeholders' (i.e., landowners) lack of experience or valuable specific knowledge into account.

The Argument for Putting Land First

Finally, one of the key claims made in this book is that land should be dealt with much earlier in the planning process of nature-based FRM. This conclusion highlights the key aspects accompanied with this approach: dealing with land ownership, the role of the facilitation of processes, how communication across disciplines matters, and the understanding of time and scale. Addressing such challenges, this volume advises fostering a more effective, more efficient, and probably a more legitimate way of implementing nature-based FRM on private land. This was proven by those cases in which conflicts of interests and values are absent or dealt with accordingly: an individual farmer decides upon the use of his land using his own resources, but this is voluntary and produces positive externalities (such as biodiversity enhancement) for others; state forest managers build retention ponds with the use of state funding on state-owned land. One of the lessons to be learned is the early engagement of landowners, planners and the public (whenever public resources are in charge) to reconcile often competing views to lock-in situations.

Probably the most important and most practical conclusion of this volume is that the book makes it abundantly clear that nature-based FRM necessitates that disciplines learn to and do communicate with each other.

Acknowledgements Open access of this chapter is funded by COST Action No. CA16209 Natural flood retention on private land, LAND4FLOOD (www.land4flood.eu), supported by COST (European Cooperation in Science and Technology).

References

Needham B, Buitelaar E, Hartmann T (2018) Planning, law and economics. The rules we make for using land, 2nd edn. Routledge

van den Brink M (2009) Rijkswaterstaat on the horns of a dilemma. Delft

Thomas Hartmann is Associate Professor at the Landscape and Spatial Planning Group of Wageningen University. He combines in his research an engineering perspective on environmental science with a socio-political perspective on FRM and land policies. He is vice-chair of the LAND4FLOOD Cost Action, speaker of the advisory board of the German Flood Competence Centre and active member of the OECD Water Governance Initiative. He is also Vice-president of the International Academic Association on Planning, Law, and Property Rights (PLPR).

Lenka Slavíková graduated from the University of Economics, Prague (public economics and policy). Currently, she serves as Associate Professor at the Faculty of Social and Economic Studies J. E. Purkyně University in Ustí nad Labem. Her long-term interest is in water and biodiversity governance with the focus on Central and Eastern European Countries. She investigates flood risk perception of different actors and financial instruments for flood recovery and mitigation.

Simon McCarthy with a background in commercial social research, undertakes teaching, training and academic social research in the Flood Hazard Research Centre at Middlesex University London. His research interests focus on the role of both public and professional social contexts in decision making, communication of risk and uncertainty and approaches to participatory interaction in flood risk and water management. Simon is appointed member of the Thematic Advisory Group on flood and coastal erosion risk management research and development for England and Wales; Department of Environment, Food & Rural Affairs, Environment Agency and Natural Resources Wales.

www.ingramcontent.com/pod-product-compliance
Lightning Source LLC
LaVergne TN
LVHW082225200926
848484LV00008B/2308
* 9 7 8 1 0 1 3 2 7 4 9 8 5 *